KT-412-999

David Berlinski wa⸱ ⸱⸱⸱
from Princeton Univ⸱⸱⸱⸱y. He has taught mathematics, philosophy and English at Stanford, Rutgers and the City University of New York, and mathematics at the University of Paris. He has been a research fellow at the International Institute for Applied Systems Analysis in Austria and the Institut des Hautes Etudes Scientifiques in France. His books – which include *The Advent of the Algorithm*, *A Tour of the Calculus* and *Newton's Gift* – have been translated into more than a dozen languages. David Berlinski now lives and works in Paris.

213 825

Infinite Ascent

A Short History
of Mathematics

DAVID BERLINSKI

NORWICH CITY COLLEGE LIBRARY		
Stock No.	213825	
Class	510·9 BER	
Cat.	SSA M	Proc. 3 WK

PHOENIX

A PHOENIX PAPERBACK

First published in Great Britain in 2006
by Weidenfeld & Nicolson
This paperback edition published in 2007
by Phoenix,
an imprint of Orion Books Ltd,
Orion House, 5 Upper St Martin's Lane,
London WC2H 9EA

1 3 5 7 9 10 8 6 4 2

Copyright © David Berlinski

The right of David Berlinski to be identified as the author
of this work has been asserted by him in accordance with the
Copyright, Designs and Patents Act 1988.

All rights reserved. No part of this publication may be
reproduced, stored in a retrieval system, or transmitted,
in any form or by any means, electronic, mechanical,
photocopying, recording or otherwise, without the prior
permission of the copyright owner.

A CIP catalogue record for this book
is available from the British Library.

ISBN-13 978-0-7538-2183-1

Printed and bound in Great Britain by
Mackays of Chatham plc, Chatham, Kent

The Orion Publishing Group's policy is to use papers
that are natural, renewable and recyclable products and
made from wood grown in sustainable forests. The logging
and manufacturing processes are expected to conform to
the environmental regulations of the country of origin.

www.orionbooks.co.uk

For Susan Ginsburg

The number of pages in this book is no more or less than infinite. None is the first page, none is the last.

—JORGE LUIS BORGES,
The Book of Sand

CONTENTS

1

NUMBER

T HE HISTORY OF MATHEMATICS begins in 532 BC, the date marking the birth of the Greek mathematician Pythagoras. Having fled the island of Samos in order to escape the tyranny of Polycrates, Pythagoras traveled to Egypt, where, like so many impressionable young Greek men, he "learned number and measure from Egyptians [and] was astonished at the wisdom of the priests." Thereafter, he settled in southern Italy; he began teaching and quickly attracted disciples. Very little is known directly of his life, except that his contemporaries considered him admirable. Nothing from his own hand remains: He has been preserved against the worm of time by the amber of various literary artifacts. Admission to the Pythagorean sect was evidently based on mathematical ability. Secrecy was enforced and dietary restrictions against beans maintained. New members were required to keep silent for a number of years, a policy that even today many teachers will find admirable, and they were expected during this time to meditate and reflect. Some members of the Pythagorean sect regarded the external world as a prison, a cave filled with flickering shadows and dull brutish shapes. Let me add to this confused but static scene the heat lightning of superb mathematical intuition.

Until the mid-twentieth century, the thesis that in mathematics as in almost everything else, the Greeks were there at first light, did not require an elaborate defense. With their forearms draped in friendship over any number of toga-clad shoulders, classicists who had spent years mastering infernal Greek declensions naturally assumed that the "Greeks were fellows of another college." The history of the Ancient Near East has come into sharper focus over the past century, great scholars poring over cuneiform tablets and recreating the life of ancient empires that had until their work been swallowed up as the impenetrable *before.* They have found re-markable things, a history before classical history, evidence that men and women have used and loved mathematics in the time be-fore time began. Neolithic ax-marks have even suggested that the origins of mathematics lie impossibly far in the past, and that men living in caves, their hairy torsos covered by vile-smelling furs, chipped the names of the numbers onto their ax handles as bison grease spattered over an open fire. And why not? Like language it-self, mathematics is an inheritance of the race.

The burden of those impossibly distant centuries now disap-pears. It is roughly six centuries before the birth of Christ. The Greeks are just about to elbow their way into all the corridors of culture. They give every indication of knowing everything and having known it all along. Yet the Babylonians already possessed a remarkably sophisticated body of mathematical knowledge. They were matchless observational astronomers, and they had brought a number of celestial phenomena under the control of precise math-ematical techniques. They were immensely clever. "I found a stone, but did not weigh it," one scribe wrote. "I then weighed out six times its weight, added two *gin,* and then added one third of one seventh, multiplied by twenty-four. I weighed it. [The result came to] one *ma-na.*" "What," the scribe now asks his oil-haired stu-dents, "was the original weight of the stone?" Mathematicians are apt to see an all-too-familiar face peeping through the problems of a Babylonian scribe—*their* face, of course, ubiquitous and always the same.

But those classicists sipping sherry in the common room of time had been right all along. The Greeks were there at first light.

The natural numbers 1, 2, 3, . . . begin at one and they go on for-ever, the mathematician's dainty dots signifying an endless pro-gression. As soon as anyone attempts to cap the natural numbers, anyone can find a way to cap the cap, say by adding one to the last natural number capped. If the numbers are infinite, they are also wonderfully various. When the great Indian prodigy Srinivasa Ramanujan lay dying in a London hospital, the cold English winters eating his lungs away now ending his life, his friend, the mathematician G. H. Hardy, paid him a visit. Paralyzed by his own reticence, Hardy could think only to blurt out the number of the taxi that had brought him to the hospital—1729, as it happens.

"I don't suppose it is a very interesting number," he added.

"Oh, no, Hardy," Ramanujan replied at once, "it is the smallest number expressible as the sum of two cubes in two different ways."

And so it is: $1729 = 1^3 + 12^3 = 9^3 + 10^3$. No smaller number has this property. The story has become famous. No one quite knows what it means, but every mathematician understands why it is told.

Like Ramanujan, the Pythagoreans were taken with the inexhaustible variety of the natural numbers, their personalities. They were fascinated by 1, 3, 6, and 10, because these numbers could be expressed geometrically as triangles composed of dots. They quite understood the importance of numbers that are divisible only by themselves and 1—the *prime* numbers such as 2, 3, 5, 7, and 11; and they may well have understood that the prime numbers are fundamental, lying like dark rubies amid the pale panoply of the ordinary numbers. They discovered that certain numbers such as 6, 28, and 496 could be expressed as the sum of their divisors. They lived in caves—I mean such is the legend—and squatting there, a pile of smooth pebbles in their laps, they saw that there are square numbers as well as triangular numbers, and amicable relationships between numbers, as when each of two numbers is the sum of the other's divisors, or when the sum of two consecutive triangular numbers such as 3 and 6 is a square number, and progressions from one series of numbers to another; and in all this, as the tallow dripped from their candles, they treated the natural numbers as if they were themselves men at play, serious but never solemn, their endless curiosity amounting at times to a form of intellectual rapture and so entirely alien to the beetle-browed scribes and accountants of the Ancient Near East, men forever plodding along the severe utilitarian axis of a commercial culture.

What did the Pythagoreans care for some pharaoh's monstrous pyramid or staring one-eyed sphinx? They were mathematicians.

Superstitious? Of course they were, but Pythagoras and the Pythagoreans were devoted to a higher spookiness. It is their distinction. With his vein-ruined hands describing circles in the smoky air, Pythagoras has come to believe in numbers, their unearthly harmonies and strange symmetries. "Number is the first principle," he affirmed, "a thing which is undefined, incomprehensible, [and] having in itself all numbers." The number one, the Pythagoreans termed the monad, and at times they seemed to suggest that the natural numbers might be subordinated to a dull grunting process by which *all* of the numbers could be generated from the monad, number creation monstrous and pullulating. "And the first principle of numbers is in substance the first monad, which is a male monad, begetting as a father all other numbers." The numbers two, three, and four enter into Pythagorean thought scent-marked from the first, the number two, because it is squat and feminine, and three, because it marks a return to the masculine, its three-tipped triangle when inverted (base up, apex down) looking very much as if a pair of wide-spread shoulders were descending toward a manly groin. The number four merits celebration—but I really have no idea why it does, except for the fact that one, two, three, *and* four sum to ten, at which point the number series topples back to one, with eleven expressed as the sum of one and ten. It is the number ten that served the Pythagoreans as the object of a sacred oath, one offered at night in the owl-hooted landscape and dedicated to "him that transmitted to our soul the *tetraktys,* which has the spring and root of ever-flowing nature."

Half-mad, I suppose, and ecstatic, Pythagorean thought offers us the chance to peer downward into the deep unconscious place where mathematics has its origins, the natural numbers seen as they must have been seen for the very first time, and that is as some powerful erotic aspect of creation itself. "Number," the Pythagoreans wrote, "is the essence of all things." Time has long scattered the Pythagoreans and canceled their sense of play, and yet the declaration that number is the essence of all things has lost none of its thrilling intellectual power. *Number?* And the *essence* of all things? Of *all* things? The Greeks heard those unearthly and mysterious words and tried to give them sense, but sand needed to sift over the monuments of antiquity before they would again enter into the mathematician's self-confident self-consciousness. When Galileo initiated the great scientific revolution of the West, writing that the Book of Nature is written in the language of mathematics, he was reconveying that Pythagorean note, those Pythagorean words.

The Pythagoreans never succeeded in explaining what they meant by claiming that number is the essence of all things. Early in the life of the sect, they conjectured that numbers might be the essence of all things because quite literally "the elements of numbers were the elements of all things." In this way, Aristotle remarks, "they constructed the whole heaven out of numbers." This view they could not sustain. Aristotle notes dryly that "it is impossible that [physical] bodies should consist of numbers," if only because physical bodies are in motion and numbers are not. At some time, the intellectual allegiances of the sect changed and the Pythagoreans began to draw a most Platonic distinction between the world revealed by the senses and the world revealed by the intellect. The literal aspect of the Pythagorean doctrine gives way. Numbers are one thing, the world of sensory objects another. But numbers still remain the essence of all things, the Pythagoreans groping their way toward the remarkable doctrine that the harmony between numbers offers a guide to the harmony between things.

"To give an example of my meaning," Aristotle remarked in describing the Pythagoreans, "inasmuch as ten seemed to be the perfect number and to embrace the whole nature of numbers, they asserted that the number of bodies moving through the heavens were ten, and when only nine were visible, for the reason just given they postulated the counter-earth as the tenth." This is neither muddied nor mystical: The inference on which the Pythagoreans relied has been championed by physicists from the seventeenth to the twenty-first centuries. It is the rock of their faith.

In the 1920s, for example, the English mathematician Paul Dirac set himself the problem of extending the Klein-Gordon field equations so that they encompassed relativistic solutions for the electron. The details of Dirac's project may safely be subordinated; what is at issue is a risky navigational maneuver in which a man sets off on a road he cannot see trusting for guidance in a road he has already seen. Dirac quickly encountered difficulties. Equations needed to be factored, as when $x^2 + 11x + 10$ is resolved into $x + 10$ and $x + 1$, and new mathematical objects were needed to accomplish this. Groping now and guessing like mad, Dirac succeeded in solving the Klein-Gordon equations, the relativistic electron appearing as a physical correlate to a mathematical object. And then Dirac noticed something odd. The solutions to the Klein-Gordon equations were split like the devil's tail. One solution corresponded to the expected electron, the solution's negative sign matching the electron's negative *charge;* but another and opposite solution seemed to correspond to the electron in all of its properties *except* charge. Lesser mathematicians might have discretely discarded this anomalous solution and carried on. Dirac ignored their advice and affirmed the existence of the positron.

He had *seen* the fork on the devil's tail. Some years later—not many, mind you—experimental physicists confirmed the existence of the positron.

Let us by all means cast out what is unwholesome in Pythagorean lore—the beans, the number mysticism, and the mumbo jumbo. There remains this: The doctrine that number is the essence of all things, passing through the prism of a thousand philosophical tracts, remains the central insight of Western science, the indispensable key of coordination. And this, too: The fact that this key opens so many locks has often been celebrated, but it has never been explained.

— —

Greek historians tell an odd little story. A ship is sailing across the Aegean Sea, the sound of the waves slapping against its wooden hull, the rowers chanting. On board are a number of mathematicians, Pythagoreans all, although why a group of mathematicians should have decided on an ocean voyage, I do not know.

Now Pythagoras is best known in the official history of mathematics for the theorem that bears his name. I am going to need that theorem close at hand, where it can do some good. Consider a right triangle whose tips are labeled *A, B,* and *C.* Distances between distant points are what they seem—*distances, and so numbers.* If distances cannot be correlated with numbers, very little remains of the grand Pythagorean proposition that number is the essence of all things. Although very simple, a right triangle—any old right triangle—is an object quivering with unsuspected numerical relationships; and in particular, Pythagoras discovered that when it comes to any old right triangle, the distances between *A* and *B,* and again between *B* and *C,* are coordinated with the distance between *A* and *C.* They are, in fact, coordinated by the simple formula: $(A - B)^2 + (B - C)^2 = (A - C)^2$. The proof that Pythagoras offered is all grunt and shove. A number squared suggests an *area* in prospect. The area of a square is, after all, the product of its length and width, and since, by definition, they are the same so far as squares go, it is the product of its length *or* its width with itself. Pythagoras thought to construct squares along each of the triangle's sides, and then by a series of geometrical adjustments—the grunt now follows—which involved shoving those squares around, he showed that the square . . .

But now that I have given away the key, the readers may follow Pythagoras through that open door. In mathematics, it is always the key that counts.

The Pythagorean theorem dooms any naive version of the Pythagorean program, the denouement taking place on board that sailing vessel just recently seen leaving port. A mathematician named Hippasus of Metapontum has just drawn a right triangle whose sides are one unit in length on the dusty surface of a ship's plank; throat cleared wetly to draw attention, he observes that by the Pythagorean theorem, the length of its diameter must correspond to the square root of two.

Now suppose, Hippasus continued, that the square root of two is a number or that it may be represented as the ratio of two numbers. In that case, $\sqrt{2} = m/n$. The steps that follow have a concision suggesting the taps of a telegraphic key:

Tap. Suppose that m/n has been reduced to its lowest common form by division.

It follows that either m and n are both odd, or that m is even and n odd, or, finally, that m is odd and n even.

Nods all around. It is a fine thing to be on board a ship.

Tap. Squaring both sides of $\sqrt{2} = m/n$, it follows again that $2 = m^2/n^2$.

Tap. Then $2n^2 = m^2$, so that m^2 is *even.*

If so, then $m = 2x$, where x is now some number. This is, after all, what it means to say that m is even.

Tap. Squaring things lavishly, it follows that $m^2 = 4x^2 = 2n^2$. . . .

My telegraphic taps now end just before the final tap; but like a newspaper announcing a great victory in headlines with details to follow on subsequent pages, this message is really complete. To get to those subsequent pages, the reader need only see that $n^2 = 2x^2$ so that . . .

But if *my* taps have come to an end, Hippasus kept right on tapping, pointing out with evident satisfaction that a contradiction had been reached, and that—*tap, tap, tap*—it consequently made no sense to suppose that the square root of two corresponds to the ratio of two numbers, and that—*tap, tap, tap*—it follows that certain distances cannot be measured by the natural numbers at all, and that—*tap, tap, tap*—

But here the story really ends. The Pythagoreans pitched Hippasus overboard where, still tapping, he perished ignominiously.

It is said that at some point in his mathematical career, Pythagoras proclaimed himself a god.

He was right to do so.

2

PROOF

MATHEMATICS IS INSIGHT AND invention and the flash of something grasped at once, but it is also something salt-cleaned and stout as a Gothic cathedral. The Pythagoreans were men of insight, and they were daring metaphysicians, too; but neither heavy lifting nor long-term construction was in their line and they were content to allow their thoughts to sparkle in the moonlight. Two centuries after the Pythagoreans trooped off, the work of salt-cleaning and cathedral-making was undertaken by the Greek mathematician Euclid. During the Middle Ages, Euclid came to be known as Euclid the Alexandrian or Euclid of Megara, but both attributions are incorrect, and stripped of the burden of two false names, Euclid has come down to us as the Euclid of the *Elements,* the book that established his immortality. This is the long view, of course, and one that has displaced many minor mathematicians between the sixth and the fourth centuries BC to the scholar's footnotes, but every cathedral has its mice.

Like Pythagoras, Euclid is largely a man of mystery, with even the dates of his birth and the city of his origin unknown. It is the Greek philosopher and mathematician Proclus who has provided the most extended commentary on Euclid's life. It amounts to only a single paragraph. "The man lived," Proclus writes, "in the time of the first Ptolemy." Euclid was thus younger, Proclus adds, than Plato's students and older than Eratosthenes and Archimedes. Ptolemy I, the ruler of Egypt and so a midget among these mighties, makes a brief ignominious appearance in the account that Proclus offers, asking "if in geometry there was any shorter way than the *Elements.*"

"There is no royal road to geometry," Euclid informed the pharaoh brusquely.

Very conscious of the importance of his subject, Euclid maintained a sideline in caustic commentaries. According to Stobaeus,

another Greek commentator, a student asked innocently enough
what profit he might gain from studying geometry. Euclid demanded that a slave give the student a few coins, "since he must
make gain out of what he learns." The coins tinkle, drop, and tum-
ble in the dust. Stobaeus, Euclid's uncomprehending student, and
that obliging slave are all alive at roughly the beginning of the
fourth century BC. And so, of course, is Euclid. He is admired, con-
sulted, respected, and talked about; he is known; he gets around,
bustling industriously. And then with a disarming indifference to
time and place, he vanishes on a shrug of eternity.

For a very long time, the *Elements* was known to every educated
man and woman, so much so that when, seven centuries after Euclid's death, a philosopher addressing a gathering of Roman intel-
lectuals asked slyly "how to construct an equilateral triangle given a
straight line," the company at once caught the reference to the very
first proposition of the *Elements*, and with the satisfaction of men
congratulating one another for being well read, broke into Greek in
order to comment on the masterpiece that had formed their character. Warm throaty chuckles all around. When, in the seventeenth
century, Isaac Newton completed his majestic *Philosophiae Natu-
ralis Principia Mathematica*, and so created the first and the greatest
of physical theories, he chose to express his thoughts in the language of Euclidean geometry, covering up as many traces of his own
mathematical inventions as he could, so great was Euclid's authority still.

The Pythagoreans had been intoxicated by the natural numbers;
Euclid was a geometer, a man proposing to impose order on the
sensuous but shifting shapes of experience. His *Elements* is a great
work of art and like all such works it serves many masters, all of
them resident in Euclid's own spacious intelligence. In the most obvious sense, the *Elements* is a textbook. It proceeds from the simple
to the complex. It is beautifully organized. It is very clear, succinct as
a knife blade. And like every good textbook, it is incomprehensible.

Euclidean geometry calls for a collaborative effort between the initiated and the unenlightened, the teacher droning, the student drowsing, until mastery of the material builds slowly in the warm space between droning and drowsing.

No matter the historical importance of high-school trots, textbooks have scant purchase on immortality. There is a treatise behind Euclid's textbook, a greater, grander book, one addressed to mathematicians, and so to men of the trade. And addressed, of course, to *us*. Now the materials of plane geometry have an existence that owes nothing to mathematics itself. There they are—the points, lines, angles, circles, triangles, squares, and squat rectangles that we encounter in ordinary life. The tabletop marks out a square; the pen leaves a dot; the ruler a line; and shadows in bright sunshine lie at certain angles to the walls or church steeples that cast them. In measuring the interior angles of various triangles, Egyptian land-surveyors certainly knew the obvious: The sum of those interior angles comes to more or less 180 degrees.

More or less, note.

Taken as a treatise, a *theory,* in fact, the *Elements* brings order to the shifting and perpetually confused detritus of experience. Practical geometry is an empirical undertaking, living and breathing and sweating in the real world where measurements are always approximate and things are fudged or smeared or jumbled up. Within Euclidean geometry points are concentrated, lines straightened, angles narrowed; idealizations are made, and some parts of experience discarded and other parts embraced. The triangle made by touching thumbs and forefingers together (in order to frame a scene, say) now disappears, replaced by the Euclidean triangle, at once perfect and controlled, a fantastic extrapolation from experience, an entry into the absolute. In the Euclidean triangle, all lines are straight, all angles crisp, and interior angles sum to precisely 180 degrees.

Precisely, note.

We are now in a position, you and I, to appreciate the third incarnation resident within the *Elements*. Just as there is a treatise behind Euclid's textbook, there is a tome *behind* his treatise, for the *Elements* is not only a book about geometry; it is, as well, a book about *how* a book about geometry should be written, and so comprises that darling of post-modernist literary studies, a meta-text.

The third tome or treatise—a disquisition on method—answers to a large and general question: How in mathematics do mathematicians achieve certainty? One answer, of course, is that they do not and that they cannot, but whether this answer is certain is far from certain, and if it is not, what then is its use? It is around a number of very similar circles that much post-modern thought accelerates without ever gaining speed. Another answer, the one offered by Euclid, and by mathematicians in every era since, is that certainty is achieved by a most peculiar method. The method that Euclid championed is the method of *proof;* and with this method, Euclid created a technique for doing mathematics and a way of being a mathematician as stylized and as demanding as the Kabuki theater.

A proof in mathematics is an argument and so falls under the controlling power of logic itself. By one of those troubling coincidences that lie littered in the history of thought, Aristotle created the discipline of formal logic at roughly the same historical moment that Euclid created his own system of geometry. Although not men of the same generation, Euclid and Aristotle stand arm in arm, linked in thought, linked in time, and linked in history. But formal logic is wider than mathematics: Its subject is inference and argument in general, and a mathematical proof is a finer, more specialized instrument than an argument in theology or the law. Not until the twentieth century would mathematics and logic, having for so long exchanged their moist breath, fuse ecstatically into the single subject of mathematical logic.

Within mathematics, a proof is an intellectual structure in which premises are conveyed to their conclusions by specific inferential steps. Assumptions in mathematics are called *axioms*, and conclusions *theorems*. This definition may be sharpened a little bit. A proof is a finite series of statements such that every statement is either an axiom or follows directly from an axiom by means of tight, narrowly defined rules. The mathematician's business is to derive theorems from his axioms; if his system has been carefully constructed, a gross cascade of theorems will flow from a collection of carefully chosen axioms. Such is the method of proof in outline, but no outline does justice to the stringency of the method or the unusual demands it places on mathematicians. A mathematical proof is like nothing else in intellectual experience, all the more reason to regard with astonishment Euclid's achievement in creating the method and simultaneously putting it to use in the *Elements*. It is rather as if he had managed to give birth to himself.

If the method of proof offers the mathematician the prospect of certainty, it is a form of certainty that is itself conditional. A proof, after all, conveys assumptions to conclusions, or axioms to theorems. If the hammer of certainty falls on the theorems, it cannot fall on the axioms with equal force.

Euclid divided his assumptions into three categories: the definitions, the axioms, and the common notions. The definitions, it must be said at once, are disappointing. There are in all twenty-three and each suggests that Euclid is attempting an intellectual task that he cannot complete. Thus a point, Euclid writes, is that which has no part, a line that which has no extent, and the extremities of a line are points. Such are the first, second, and third of Euclid's definitions. The criticism that logicians make is that these definitions are either circular or that they commence an unhappy regress. To know that a point is that which has no part is hardly helpful if having no part is defined in terms of being *point*-like; and if not defined in such terms, then in what further terms? Caught between the circle and the regress, modern texts in geometry simply *list* their undefined terms, making no attempt to endow them with meaning. Other terms are defined explicitly by reference to the undefined terms. This way of doing things is wholesome and correct, the chain of definitions backing up to dead-end finally at terms whose meaning is either assumed or ignored.

Euclid's common notions, on the other hand, are sensible enough. Falling into the category of the invaluable intellectual bromide, they give no offense. Let me list them all:

1. Things that are equal to the same thing are also equal to one another.
2. If equals be added to equals, the wholes are equal.
3. If equals be subtracted from equals, the remainders are equal.
4. Things that coincide with one another are equal to one another.
5. The whole is greater than the part.

Euclid termed 1–5 *common* notions because he felt that in some sense they must be a part of any mathematical system dealing with geometry. Modern logicians would assign these notions to logic itself, but no matter their natural home, no one is apt to provoke intellectual indignation by insisting that the whole is greater than its parts.

There remain the axioms, the germinating seed of Euclidean geometry. The axioms must meet two constraints: They must be rich enough so that everything important about the world of geometry may be derived from them; and they must be sufficiently self-evident so that they may be accepted without argument. Euclid's axioms are not perfect. There is a worm hidden in them. But judged by the standards of his predecessors, the system they make possible is not only remarkable but unprecedented, Euclid the greatest of ancient system builders because the first.

There are in all only five axioms needed to make possible the creation of the Euclidean world. The first three are constructive in their import: They affirm that something can be made; they are enabling in their effect. Let the following be postulated, Euclid writes, that it is always possible:

1. To draw a straight line from any point to any point;
2. To produce a finite straight line continuously in a straight line;
3. To describe a circle with any center and distance.

Axioms 1–3 have a simple, easily grasped nature. Thus Axiom 1: Where there are two points, there is one line. And thus Axiom 2: Where there is a straight line, there is a longer straight line, and so without end. And Axiom 3: Circles for the asking.

The fourth axiom is a declaration covering under the aspect of equality *all* right triangles wherever they may be found in space:

4. All right angles are equal to one another.

There remains the fifth axiom of Euclid's system, and with the fifth, that worm. It is a worm that may now be seen wriggling in words due to the eighteenth-century Scottish mathematician John Playfair:

5. Through a point outside a given line, one and only one line may be drawn parallel to that line.

Whether in Euclid's original formulation or in Playfair's, these are words that have haunted the mathematical community. The axiom that they express is a vital part of Euclid's system, a load-bearing structure. And surely it seems plausible. One straight line; one exterior point; and only *one* line through the point and parallel to the given line. Yet the picture corresponding to the parallel postulate does not cancel a sense of mathematical unease. In some very obscure way, the axiom contains an assumption that it does not entirely succeed in conveying. Parallel lines and a point in space—clear enough. And the picture that results—clear enough as well. There is yet something odd and unresolved about the picture of those jaunty parallel lines, its visual plausibility depending entirely on the assumption that the space in which they are embedded is *flat*. If that assumption is canceled or otherwise modified, the picture at once loses its initial plausibility, space itself acquiring the power to droop in strange ways . . .

And with the mathematician's dots now making an unwelcome appearance, we realize that Euclid's fifth postulate does not quite have the same hold on the organs of the obvious that his other four assumptions do. It is this that has suggested to mathematicians throughout the ages that the fifth axiom is no axiom at all, but, instead, a *theorem* of the system. For more than twenty-two centuries, mathematicians attempted to make good this insight, demonstrating in a hundred cobwebby and confused papers, whether in Latin, Greek, Italian, French, or German, that Euclid's fifth axiom could be *derived* from Euclid's other axioms. In the end, every such demonstration seemed to assume precisely the point at issue, either obviously, as when hearty amateurs had a go at things, or in some monstrously subtle way, with excellent mathematicians following a cunning series of circular steps inevitably conveying them back to the parallel postulate itself.

And this is something that Euclid knew, understood, and found deeply troubling. Never once did he propose a proof of his parallel postulate, his superb intuition having perhaps seen straight through the tangled trail from one fallacious proof to another to gather itself at the incredible conclusion that the parallel postulate was—no, not a theorem, in that way lies madness—but in fact . . .

What a wonderful instrument trailing dots turns out to be, with ever so many literary techniques abbreviated in their dainty drumbeat: foreshortening, far shadowing, fast forwarding; they are an invitation, those dots, a guide to romance, a tease, a sign of the imponderables to come.

In writing the *Elements,* Euclid was not a man to trifle with warmups. The book opens with a bang of brusqueness. There are no preliminaries. Something is to be demonstrated. Given a line, Proposition I affirms—the very first sentence of the book, mind you—it is always possible to construct an equilateral triangle on that line. To construct—meaning, to *create.* A new geometrical object now arises from Euclid's definitions and axioms. The proof marches in short, severe Roman-troop steps. And then all at once it stops. The thing to be demonstrated has been demonstrated. The concision is almost unbearable.

And thence to those steps.

1. Let AB be the given finite straight line.
2. With center A and distance AB, let the circle BCD be described.

Axiom 3.

3. With center B . . .

Axiom 3 again.

4. From the point C at which the circles intersect, draw a straight line to the points A and B.

Axiom 1.

5. But $AC = AB$. . . .

And with only five inferences completed, the architecture of the proof is evident. Taking a trip of three steps around that triangle, Euclid shows methodically that its sides are all equal: AC and AB, and BC and BA, and CA and CB, which is nothing more than AC and AB.

So the *Elements* begins, and so it continues, passing from very simple propositions to propositions that are quite complex, and passing as well from what is obvious to what is entirely unsuspected. Over the long centuries in which Euclid's *Elements* have been the cynosure of every mathematician's eye, its secrets have been uncovered and its dark places flooded with light. By the nineteenth century, the system held few surprises. And yet like the sturdy old system that it is, Euclidean geometry is still capable of sending out a few resplendent springtime shoots. In the late nineteenth century, for example, the Anglo-American geometer Frank Morley discovered and demonstrated that the angle trisectors of *any* Euclidean triangle form an interior equilateral triangle. This is an exquisite result, one reached twenty-three centuries after Euclid's death, striking evidence that the method of proof is also an instrument of discovery.

At some time in the seventeenth century, the French mathematician Pierre de Fermat asked whether the equation $x^n + y^n = z^n$ could be solved by means of distinct integers. At $n = 1$, the question is trivial; and at $n = 2$, obvious, the Pythagorean triplets three, four, and five very elegantly satisfying the equation $x^2 + y^2 = z^2$. Three squared plus four squared just *is* five squared. As exponents mount past two, polarities reverse. Search as he might, Fermat could discover *no* Pythagorean triplets x, y, and z such that $x^3 + y^3 = z^3$. "It is impossible," he wrote in the margins of the Diophantine treatise *Arithmetica*, "for a cube to be written as the sum of two cubes." He then took the step that would immortalize his name. He generalized his observation. What cannot be done for third powers cannot be done at all, no matter the power, no matter the search. It is *impossible* "for any number which is a power greater than the second to be written as a sum of two like powers."

Fermat believed that he had discovered a marvelous proof of his own conjecture, and within the margins of his own paper noted sadly that the margins were too small to contain it.

Very good mathematicians were intrigued and often obsessed. Amateurs and cranks, all of them curiously aware of my e-mail address, busied themselves with crackpot proofs, some of them fiendishly ingenious. For more than three centuries the conjecture remained unyielding. And then in 1993, the English mathematician Andrew Wiles announced a proof, one retrospectively validating Fermat. The old boy had been right after all. Wiles' proof ran to more than two hundred pages and it made use of an immense body of modern mathematics. A first version, announced in a very dramatic setting at Oxford University, contained an error. The proof required revision. But then everything came right.

Although his paper addresses an old problem, it is completely an exercise in the most modern mathematics. It is, in fact, *hyper-modern*. And yet there are features of this paper that are old rather than new. The proof is written in the service of a Pythagorean obsession, all of the old half-mad Greek Pythagorean voices gathering again to speak and sing. Fermat's conjecture plays over the simplest properties of the numbers and the question that it raises, whether for any *n* greater than two there *are* numbers *x, y,* and *z* such that $x^n + y^n = z^n$, is so plainspoken and seems so close to the bone of intuition that it almost invites a spontaneous declaration. Yet the proof of Fermat's last theorem lies quite beyond intuition, correcting the Pythagoreans in their madness by showing that intuition in mathematics must always be structurally supported.

Whereupon Euclid coughs discreetly in the night. Despite its magnificent complexity, and despite its ferociously hyper-modern symbolism, Wiles' paper that is organized in accordance with precisely the architectural plan on display in the half dozen or so lines that Euclid required to prove the first proposition of the *Elements*. Something in both cases is to be demonstrated and so made captive to the method of proof. Some things in both cases have been assumed. Some common notions are in both cases taken for granted. Inferences control the flow of thought. The cathedral of mathematics has increased in size but not in its inner nature. In no other human subject, I suspect, has so much changed and so much stayed the same.

3

ANALYTIC GEOMETRY

T O THE GREEKS ALL credit. All credit, and then there is silence. The stolid Romans, who had conquered the Greeks and then conquered the world, were brilliant military men. They had a genius for politics and propaganda; and they were gifted in the law, medicine, and sanitary engineering, three disciplines that have done more for human happiness, I suspect, than any other human undertaking. But the Romans possessed no mathematical gift whatsoever, their incompetence as striking as it would have been had classical Greek culture given out directly to modern-day Rwanda or the Sudan. Mathematical curiosity died in the Roman Empire and it stayed dead in the Christian West for more than one thousand years. There were great theologians and philosophers, to be sure: the Church fathers, the Venerable Bede, Anselm, Abélard, Albertus Magnus, Thomas Aquinas, Duns Scotus, William of Occam; but no one on fire with the Pythagorean rapture, only men prepared indifferently to sift its ashes. In the great Moslem Empire that from the eighth century AD to the middle of the thirteenth century stretched from Spain in the west to the borders of India in the east, things were otherwise. Arabic was the language of literate men and women, the suave and supple intermediary between Greek and Latin antiquity and medieval Europe, and the perfumed city of Baghdad was their dimpled pleasure pool, the center of the Arab archipelago. After defeating his brother in battle and so securing power, the Caliph al-Ma'mun created the House of Wisdom in the early part of the ninth century; refreshed by conquest, he invited mathematicians, astronomers, astrologers, poets, and translators to mingle in the cool marble of its corridors. Arab mathematicians invented a flexible notational system for the natural numbers—the decimal system still in use; they learned to solve quadratic equations. They carried out a flirtation with the negative numbers and incommensurable magnitudes, one of those dusky desert duets in which owing to a remarkable amount of unnecessary clothing neither party quite knows what the other is really like. They were daring. Writing during the ninth century, the Moslem Renaissance covering all Baghdad in its aureate and ochroid glow, al-Khwarizmi handled square roots and powers with an easy familiarity; his disciple Abu Kamil had a way with higher powers; using essentially modern methods (subtract from both sides, factor, hope for the best), he was able to solve a number of quadratic equations. Wise far-seeing Abu. And there is the Omar Khayyam of the *Rubiyat*, a Persian among Arabs, and so a songbird among sparrows, a mathematician of note, occupied with the solution of cubic equations, his lyrical intelligence finding in algebra the anodyne against time that time had long withheld.

And this, too, is noteworthy: An immense body of Arabic scientific literature remains unread and unstudied. It is possible that scholars centuries from now will refer respectfully to the *Arabic* Newton, some ferocious far-seeing intelligence whose vexed spirit remained sputtering with indignation during all the long centuries in which *his* masterpiece lay buried in dust-covered stacks or on the shelves of some antiquarian's bookstore.

Whatever the gems that scholars have not seen, the fact remains that studying the history of mathematics *today,* the historian may skip from the end of the Greek era to the beginning of the modern era without ever troubling his scholarly conscience.

And needless to say, of course, if he is about to skip, why should we linger?

It is 1600. Adieu scholastics, steeples, and scribes. The great age of mathematics is about to commence. At the beginning of the seventeenth century, the mathematician Marin Mersenne could envisage all of mathematics contained within the corners of a single library shelf, the collection commemorating a few antique starbursts such as Euclid, Appolonius and Archimedes, and a number of late Renaissance shooting stars—Cardano, Torricelli, and Bombelli. Mersenne was himself a polished old number hand and that shelf contains a treatise or two about the natural numbers. There is a discreet pause, and then Mersenne mentions the number zero, the invention of an Indian mathematician, and the negative numbers −1, −2, −3, −4, . . . , the mathematician's dots abbreviating another progression, this one going *backward* toward the black badlands of infinity. The mathematicians of the Renaissance were easily spooked, Mersenne confides, both Nicolas Chuquet and Michael Stifel regarding the negative numbers as if they were the devil's playthings.

Mersenne has very little to say about the *real* numbers. He mumbles when the matter comes up. The Greeks knew that the square root of two could not be correlated with the natural numbers or with the ratio of such numbers. A man had died to prove the point. More than fifteen hundred years later, dust and grease still covered the subject, the first to obscure vision and the second to prevent traction. There are the natural numbers; there is zero, and the negative numbers; and there are the fractions. Fractions give rise to repeating decimals, as when 1/7 is represented by 0.142857 . . . with never a change at the decimal's tail. Then there is this business about the square root of two. It cannot be represented as an ordinary fraction. But the number, whatever it is, can be *approximated* by a number dragging an endless decimal. This is something that the Babylonians knew. The square root of two is, after all, more than 1 and less than 2, and then more than 1.1 and less than 1.9, . . . the approximation getting better and better as those dots and that process go on and on. Modern calculators return 1.4142135 . . . when the square root of two is requested. Computers do far better. But no matter how far out the approximation runs, it never settles into a pattern. The decimal expansion is irregular. It looks for all the world as if it were random. It is all very confusing. Like other mathematicians of his era, Mersenne has nothing of interest to say about these numbers.

And then popping up as curiosities, and so completing the impression that the numbers are very odd, there are within Mersenne's ken numbers answering to equations such as $x^2 = -1$. Rafael Bombelli had placed his hand on what are now called complex numbers; he saw just how they might be manipulated. He had discovered pure gold, but then in a burst of sunny Italian carelessness, he somehow let things lapse.

Mañana.

René Descartes was born in 1596 and died fifty-four years later. The greatest of early modern philosophers *and* the greatest of early modern mathematicians, Descartes is unusual in the degree to which his thoughts were guided by a program, a settled way of looking at things. It is a program that had its origins in disappointment. "I had when younger," Descartes remarks, "studied logic . . . and geometrical analysis and algebra." To little avail, apparently. The syllogisms of logic, he observes sadly, "serve to explain to another what one already knows." Geometry and algebra are hardly an improvement. Geometry is "so limited to the consideration of figures that it cannot exercise the understanding without greatly fatiguing the imagination." And by the same token, "algebra is so limited to certain rules and certain numbers that it has become an obscure art which perplexes the mind rather than a science which educates it." Descartes was plainly a man ill disposed to accept intellectual gifts.

The method that Descartes commended is often called the method of doubt; it has the virtues of any program of self-improvement. *Manage the problem; be careful; accept nothing on faith, look for clear and distinct ideas, trust in yourself.* Descartes was by no means a skeptic, a man prepared to empty his thoughts by placing their objects in jeopardy, but he recognized that the familiar material world of trees and trellises, sunshine and shade, lies at the end of a complicated and desperately fragile inferential trail, with even the most obvious of physical facts—*the rose is red*— open to epistemological corrosion. *The rose? Red? Really?* How do you know—yes, *you*—that you are not dreaming, mistaken in perception, flawed in judgment, or simply the victim of a cruel cosmic tangle, your senses deranged by a demon? These questions have passed into the universal curriculum of mankind, where every now and then, as philosophy teachers well know, they are still able to convey sensitive students to the edge of madness.

As the external world recedes, the mind returns to itself. *I think, therefore I am.* Engaged by and with itself, the mind is proof against doubt, the distinction between the way things seem and the way they are vanishing, seeming and being blessedly annealed. And thereafter Descartes translates these observations into a metaphysical system. The universe is divided in its nature. There is the world of matter and the world of mind. It is the individual, naked in his thoughts and alone, who must learn how to represent the external world. One key to that representation, Descartes sensed, may be found in mathematics, for elementary mathematics represents an aspect of the mind's conversation with itself, and so is something that shares in the mind's general proof against doubt.

Descartes received his early education from the Jesuits at the academy of La Flèche, and he apparently persuaded his teachers that his was a sensitive disposition. He was allowed to sleep late and thereafter became an inveterate valetudinarian and a consummate whinger. Moving to Paris in 1612, he encountered his school friend Marin Mersenne, the two young men talking late into the endless night, their wooden boot-heels clacking along the Paris streets (not far from where I now live). Descartes then joined Dutch military forces under William of Orange and later volunteered to serve in the Count de Bucquoy's Bavarian army at the beginning of the Thirty Years' War. Whether his commanders indulged his desire to sleep late each morning and not be disturbed in general, Descartes does not say. On the night of November 19, 1619, encamped in some muddy field by the Danube, with military pennants fluttering in the cold wind and the tent walls sullenly flapping, he was vouchsafed a dream in which the secrets of both philosophy and mathematics were revealed.

The book that most completely advanced Descartes' agenda and expressed his dream is entitled *Discours de la méthode pour bien conduire sa raison et chercher la verité dans les sciences—The Discourse on Method;* analytic geometry appears in an appendix entitled *La Géométrie.* It is a book that suggests more than it contains and it is written by a mathematician unaware of the treasures he was in the process of uncovering, for the curious fact is that the leading ideas of analytic geometry are not expressed explicitly in the work that created analytic geometry, so that in reading the book in the original, mathematicians very often come to wonder how its author could have seen so much while saying so little.

At some time during the thirteenth century, Gervasius of Tilbury, a well-educated English nobleman, conceived the idea of creating a map of the world. Gervasius was both a scholar and an historian. He got around, traveling as far afield as Bologna for his education, and then spending time at the court of William II in Sicily. He may be spotted after that in Burgundy, serving as an adviser to Otto IV just before Otto's defeat at the Battle of Bouvines. Thereafter, the trail grows tangled as Otto, his dreams of world conquest at an end, settled into lower Saxony, the ever-helpful Gervasius behind him. A book survives—the *Otia Imperialia*—which Gervasius wrote apparently to amuse the king.

And in a well-preserved copy, the great map survives as well—the *Ebstorf Mappamundi.*

Oriented toward the east (from the perspective of Hanover, where it was originally located), the map depicts the world as it is enfolded tenderly by Christ, whose (remarkably small) head is located in the Garden of Eden, which the mapmakers have placed in India. The rest of the world follows the contours of Christ's body, his right hand draped carelessly over northern Albania, and his left over southern Africa. At the very center of the map and so the place where all roads converge is the city of Jerusalem. The tombs of all the apostles are depicted on the map, together with the chief Roman churches. No distances; no ruins or runes; no geographical landmarks. A number of animals are resident on the map, and so, too, men whose sperm-whale torsos end in narrow dog heads, a few ruddy giants rising like stalks from the margins of various savage empires.

By the seventeenth century, road maps, military maps, and maps of coastal seas had accumulated in all the chests and drawers of Europe. In Prague, the infinitely exuberant Tycho Brahe, his metal nose plate elegantly concealing a dueling scar, had begun to create a map of the heavens; and in the low countries, Hieronymus Bosch had completed a depiction of Hell, one so detailed that it, too, counts as a map of sorts, a guide to all of Europe's rotting monsters. It is this world of draftsmen and cartographers that René Descartes now enters, and like everyone else, he has a map in hand, a representation of the world.

The Cartesian map is written not on parchment or on paper, nor for that matter is it etched in metal. It is and it remains a mental map, one inscribed on the Euclidean plane. It can be seen, but only with the same eye capable of seeing points without dimension, straight lines without width, and triangles that do not bend, sag, or crumple. Its fundamental plan is simple. The plane is first bisected by two straight lines. Their point of intersection is labeled 0. This is the origin of the world map and so its new Jerusalem. The numbers depart Jerusalem in four directions along four straight lines: to the right, where they increase positively $1, 2, 3, \ldots$; and to the left, where they decrease negatively $-1, -2, -3, \ldots$; and then up and down in precisely the same fashion. A plane organized in this way is known as a Cartesian coordinate system (Figure 3.1), even though Descartes did not himself talk of coordinate systems in *La Géométrie.* Or anywhere else, for that matter.

It is difficult at first to see the profound resemblance between the Ebstorf world map and a Cartesian coordinate system, but

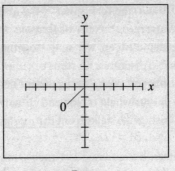

Fig. 3.1

maps, wherever they are found, share a common feature: They express far more than they contain, and so share in (and exhibit) the paradox of representation itself. A Confucian mandarin looking at the Ebstorf world map might see nothing more than a strange, narrow-headed figure, his arms flopping over colored regions of space, animals in residence here and there, and dog-headed men loitering about. Gervasius of Tilbury, on the other hand, saw a representation of the world beyond the world, countries and continents subordinated to the Christian drama of sin, suffering, and redemption. In regarding the crossed and martyred arms of Descartes' coordinate system, we, too, *we innocents,* at first see what is there without seeing what it represents. Descartes saw beyond what can be seen, precisely the reason he is regarded as a great mathematician.

It is René Descartes who is now at your elbow; with the full force of his powerful if somewhat sullen intelligence, he is offering you the keys to *his* kingdom.

Points emerge first.

With numbers inscribed on the axes of a coordinate system, any point in the Euclidean plane may be represented by two numbers. Call them x and y. These are the coordinates of the point. Each marks a certain distance along the x- and the y-axis. The point itself lies at the intersection of two straight lines, one moving vertically from x, the other moving horizontally from y.

Distance is next.

If A and B are any two points in the plane, with the coordinates (x_1, y_1) and (x_2, y_2), the distance D between them is:

$$D(A, B) = \sqrt{(x_2 - x_1)^2 + (y_2 - y_1)^2}$$

The formula should inspire the pleasure of friendship renewed, for it follows from the Pythagorean theorem. Squaring is in force in the formula, by the way, in order to get rid of negative numbers, and the subscripts on both x and y are bookkeeping devices, a way of keeping things distinct.

With the introduction of numbers, points, and a formula for distance, the first step in a great drama of identification begins. If points correspond to pairs of numbers, there is no reason that geometrical figures more complicated than points—lines and curves—might not correspond to mathematical objects more complicated than pairs of numbers, so that with the mathematician ascending step by step, the geometrical world becomes coordinated with the world of numbers.

Straight lines now follow.

A straight line within the Euclidean plane is simply a straight line, of no address and in no way distinguished from any other straight line. There are thousands of them just lying around. But given the struts of a Cartesian coordinate system, those shiftless straight lines straighten themselves out and acquire the dignity of a fixed identity. Two numbers are again required. The first expresses the way in which the line addresses the axes of a coordinate system, its angle of inclination or slope. This can be represented as a ratio $m = (y_2 - y_1)/(x_2 - x_1)$. The second—a buoyant b in most books—marks the place that the straight line crosses the y-axis and so signifies its point of intersection. Symbols are combined in a single equation in two unknown variables, $y = mx + b$. As the mathematician runs through the values of x, values for y pop obediently out of the equation's other side, the line—this line, *our* line—consisting of all those points x and y that satisfy $y = mx + b$. The line is now bound, its identity inexpugnable. It is *that* line and no other.

Equations in two variables—two *indeterminates* in the language of the sixteenth century—correspond to the locus of points making up a curve. This is the leading mathematical idea of analytic geometry and if Descartes properly perceived its importance, he very successfully concealed his enthusiasm. Curves in the plane? Their loci? "In every such case," he writes, "an equation can be obtained containing two unknown quantities." Frowning, he then occupied himself elsewhere, chiefly with a difficult problem posed originally by the Greek geometer Pappus. Yet the coordination suggested between curves in the plane and various equations is a subtle and intricate achievement, one that Descartes realized imperfectly but one that he did realize.

By the sixteenth century, mathematicians had already drawn a distinction between parameters and variables. Parameters are fixed, variables variable. The letters m and b in the equation of the straight line thus denote specific numbers. They do not vary. They stand their ground. *Which* numbers they might happen to denote is of little interest. The variables x and y, on the other hand, take different numerical values in one and the same equation, and so function rather as if they were pronouns in English.

Descartes realized, perhaps unconsciously, that the correspondence between equations and *curves* in the Euclidean plane turned on an intricate play between the form of an equation, its variables, and its parameters. This is new ground. A number of very familiar Euclidean shapes now find themselves controlled from *beyond* the plane by an algebraic alembic, the key to the coordinate system (just as the gospels provide the key to the *Ebstorf Mappamundi*). A general first-degree equation has the form $Ax + By + C = 0$. The equation has in x and y two variables; and it has in A, B, and C three parameters, temporary workers, there for the duration. In addition and multiplication, the equation specifies two mathematical operations. After those operations have been conducted on the left side of the equation, the result, the equation itself affirms, is equal to zero, striking evidence that in mathematics, as in life, much often comes to nothing. This particular equation is known among mathematicians as the general equation of the line; it is known as well as a *linear* equation, the word *linear* in wide and unhappy currency among political scientists, sociologists, and romantic counselors much concerned to advise their clients to avoid linear thinking in their relationships. Here is the real meaning of the word, one tied precisely to a certain set of symbols. Straight lines are straight and the equations that describe them are linear. Romance has nothing to do with it.

Whatever the straight lines, the distance formula can be used almost at once to derive an equation governing a circle, inasmuch as a circle in the plane consists of the set of points that are all at an equal distance R from the circle's center C. If the coordinates at C are a and b, then by a single application of the distance formula, the circle's radius R is simply the square root of $(x - a)^2 + (y - b)^2$, and the circle is now confined in a cage of symbols.

If a first-degree equation governs the straight line in the plane, a second-degree equation—an exponent in the cockpit—handles a handful of more sensuous curves. There is the equation of the parabola $y = ax^2 + bx + c$; and the ellipse, $x^2/a^2 + y^2/b^2 = 1$; and the hyperbola, $x^2/a^2 - y^2/b^2 = 1$.

These equations may be amalgamated into a single second-degree equation: $Ax^2 + Bxy + Cy^2 + Dx + Ey + F = 0$. There are in this equation two variables, x and y, and there are as well six parameters, A, B, C, D, E, and F. The equation is quadratic in x and y.

A more considerable cartographic vista now opens. Within

geometry, a cone is just what it seems to be: a well-known shape in three dimensions. By slicing the cone in various ways, the Greeks discovered to their satisfaction, a variety of two-dimensional Euclidean points, lines, and curves could be coaxed from the cone itself. Such are the conic sections, long a staple of high-school geometry.

The general second-degree equation, Descartes determined, corresponds to a conic section. What is more, $Ax^2 + Bxy + Cy^2 + Dx + Ey + F = 0$ describes

1. a parabola if $B^2 - 4AC = 0$;
2. an ellipse if $B^2 - 4AC$ is less than 0; and
3. a hyperbola if $B^2 - 4AC$ is greater than 0.

The expression $B^2 - 4AC$ is the discriminant of these equations; it stays the same no matter how the axes of a Cartesian coordinate system might be rotated; and so it comprises the first great invariant of modern mathematics.

The equation $Ax^2 + Bxy + Cy^2 + Dx + Ey + F = 0$ rather resembles the Everyman logo that used to appear affixed to certain books published years ago in central Europe, all overcoat, indistinct features, and shuffling tread. It is perhaps not surprising to see in this mathematical Everyman a *general* connection to plane curves. But to see in the jumble of variables and parameters a connection to very specific, brightly individuated plane curves such as the parabola, hyperbola, and ellipse—*that* required a mathematician's eye. The discriminant of the equation, $B^2 - 4AC$, which controls the correspondence, is not present in the equation itself, so that just as Picasso painting Gertrude Stein could see the massive woman she would become from the rather uninteresting woman she was, the mathematician can see contingencies and conditions in those otherwise unrevealing symbols that only emerge later.

Descartes' theorem is in this sense an achievement in intellectual refinement; it invites the eye to linger, and it suggests for perhaps the first time that what can be said about mathematical objects is more interesting than the objects themselves. In the appreciation of a work of art, the amateur appreciates the subject, but the connoisseur admires the *painting*.

Analytic geometry of the seventeenth century is destined to grow great, some lines of development already immanent in *La Géométrie*, others in the work of Girard Desargues, Pierre de Fermat, and Blaise Pascal. Two-dimensional analytic geometry is easily promoted to three-dimensional analytic geometry, with the third coordinate axis rising from the page like a priapic stalk.

The coordination between certain equations and certain structures in space proceeds directly up the dimensional chain. The fourth-order equation $V = x^2y + y^4/4$, for example, describes an undulating surface in a three-dimensional space.

Analytical geometry may be conducted in four dimensions, if need be, and although the results cannot easily be seen—let us be honest: They cannot be seen at all—the analysis is much the same.

When Andrew Wiles offered his proof of Fermat's conjecture, he used an immense array of tools, but at the very center of his proof a tingling trail led backward to Descartes, for what he had succeeded in proving was the Taniyama-Shimura conjecture, a thesis about elliptical *equations* and modular *forms*, one that in the complexity of its formulation hid that old, shrub-covered trail between the form of the discriminant and various curves in the plane.

Growing great in one direction, analytic geometry also grew great in another. Bavarian artillery officers certainly knew what happens when one of their cumbersome pieces lofted a cannonball into the air: The thing went up and then it went down, its parabolic arc ending in some awful splotch. Artillery tables in common use gave them rules of thumb for calculating trajectories. A cannonball in motion is a moving point—*no?* And its trajectory is a curve—*no again?* And in particular, it is a parabola—*no a third time?* Yes it is. But a parabola has a specific shape. It consists of all points satisfying a specific equation. By means of the equation, the cannonball's trajectory in space can be *completely* specified, the equation yielding one point after another, the accumulating points yielding the curve, the curve yielding the trajectory. A shape in space has given way to an analytic formula, a verbal contrivance. And with this insight, the first step has been taken in a vast, far-reaching project that will in the end bring all forms of continuous motion, the cannonball *and* the rotation of the planets in the night sky, under the control of a numerical apparatus.

Was it Pythagoras who remarked that number is the essence of all things?

Or was it obvious all along?

No one ever observed that Descartes was warmhearted; he was neither kind to animals nor fond of children, points, many mathematicians will observe, that are in his favor. A beloved but illegitimate daughter died in her young adulthood. A very curious story suggests that Descartes fashioned an automaton in her likeness and carried it around Europe with him, propping the device up in various hotel rooms, and pouring out his heart to the horrible thing. Almost the same story is told of Albertus Magnus, Thomas Aquinas finally destroying *his* automaton in a fit of deep indignation. Although affected by grief, Descartes remained unmarried, indifferent apparently to women, his manner distant and his nature cold.

In 1628, Descartes moved to Holland, a vibrant and tolerant center of science and art. On learning that the Inquisition had for every good reason placed Galileo under house arrest—the man was incapable of keeping his mouth shut or his thoughts to himself—Descartes determined to keep private views that might provoke clerical hostility or even philosophical disagreement. Descartes published his *Meditations on First Philosophy* in 1641, and almost at once the book was assigned to students in philosophy. It has the great virtue as a text of being easy to read and difficult to understand, thus satisfying the needs of both students and teachers. Descartes' *Principia philosophiae*, published three years later, expresses most completely his mature philosophy. A prophetic work, the *Principia philosophiae* represents an ambitious attempt to explain the universe by means purely of the forces that sweep through matter. Action, Descartes believed, requires contact, as when one object hits another, and the constraint of contact rules out action at a distance. If this is so, the universe must be filled either with material objects or with structures capable of placing material objects in contact with one another. Such are the Cartesian vortices. In Descartes' view, it is the various vortices, whose effects resemble on a grand scale the spiral sweep of bathwater swirling an errant bar of soap drainward, that brings about the required contact. Isaac Newton demolished Cartesian physics with all the immense power of his genius. He did nothing to assuage the Cartesian scruple about action at a distance, and having with a snort of derision dismissed the Cartesian vortices, he replaced them with the force of universal gravitation, which acts at a distance and throughout the whole of space.

In 1649, Descartes was fifty-three; he was at the height of his powers. Hearing of his competence, Queen Christina of Sweden persuaded him to attend to her intellect by accepting a position as a royal tutor. The queen apparently turned Descartes' head in the way in which most women turn any man's head, and that is by shameless flattery. Descartes moved to Stockholm where he was dismayed to discover that the queen, a squat young Amazon, expected instruction to be given early in the morning. Descartes was forced to oblige and forsaking his warm bed, trudged out into the Swedish dawn, day after day, until with his own principles of self-interest held in abeyance by royal command, he caught cold and died.

4

THE CALCULUS

THERE NOW OCCURS A reverberating sonic *boom!* in the history of thought. Before the discovery of the calculus, mathematics had been a discipline of great interest; afterward, it became a discipline of great power. Only the advent of the algorithm in the twentieth century has represented a mathematical idea of comparable influence. The calculus and the algorithm are the two leading ideas of Western science.

Within the strict confines of the definitions, theorems, and proofs of mathematics, the calculus reaches its apogee in its fundamental theorem. The theorem does what all great theorems do: It shows that two things thought to be distinct are profoundly connected. But the calculus is also the indispensable instrument required by the physical sciences, and the physical sciences have without exception followed the intellectual model established by the calculus. It is the heart and soul of their method. This suggests—does it not?—that mathematics is hardly a matter of mathematics, and its theorems are not simply statements about the mathematician's mind. They are the reflection of the living truth.

The calculus is a theory of continuous change—processes that move smoothly and that do not stop, jerk, interrupt themselves, or hurtle over gaps in space and time. The supreme example of a continuous process in nature is represented by the motion of the planets in the night sky as without pause they sweep around the sun in elliptical orbits; but human consciousness is also continuous, the division of experience into separate aspects always coordinated by some underlying form of unity, one that we can barely identify and that we can describe only by calling it continuous.

Many mathematicians had a hand in the development of the calculus—Gilles Personne de Roberval, Pierre de Fermat, Isaac Barrow, Bonaventura Cavalieri, John Wallis—and every one has acquired a contemporary scholar willing to insist that his boy had seen it all along; but it is Gottfried Leibniz and Isaac Newton who are the most closely associated with the pause between heartbeats when everything changed. It is the second half of the seventeenth century. Shall we say roughly 1680 or so?

Time has long since promoted both Newton and Leibniz into the pantheon of the great explorers. Like two immense polar bears, they remain forever frozen on the tundra of time. "He stands before us," Einstein remarked in commemorating Newton, "strong, certain and alone." These are beautifully chosen words, and they succeed in capturing the Newton of myth and the Newton of historical memory. It is rather more difficult to imagine Leibniz standing alone, or ever standing silently, if only because he was the most gregarious of men, bouncing urgently in his coach across all the roads of seventeenth-century Europe and exchanging letters with more than six hundred correspondents, some as far away as China. The man was inherently gabby, his personality controlled by an urgent need to communicate his thoughts, which endlessly overflowed their natural cistern. He comprised a one-man multitude, but the men inside that multitude were all of them outstanding, and the light of their cumulative luster quite matched Newton's single, monstrously pulsing sun.

An object—a ball, a bullet, or a ballerina—is in motion. It rises into the air, reaches an apogee, and then descends. Abstraction and idealization are now at work. That object in motion is first stripped naked, its weight, heft, mass, color, and complaint that her partner is *always vunting spotlight* discarded, so that that ballerina lumbering into space is replaced by her essence, which is change in place.

By means of mathematical meiosis, change in place divides into change in distance *and* change in time. This division places the ballerina's moving arc within reach of familiar mathematical concepts, for both distance and time may be represented by *numbers,* and then depicted on the axes of a Cartesian coordinate system.

Time is running off to the right, distance is going up. The purely verbal description of the ballerina's trajectory—*squat, lift, grunt, up, down*—gives way now to a mathematical description in which distance is represented as a *function* of time. This Galileo had already made clear in his *Two Sciences* of 1638 when he observed that the distance covered by a freely falling object is proportional to the square of the elapsed time—$16t^2$. The relationship between distance and time is very naturally expressed as a curve inscribed on a coordinate system. There is time going off on one axis, and distance going up along another. Plotting the second against the first results in a curve, the very one that now displaces the real world's ballerina in favor of one of her essences.

If distance and time are at hand, so, too, speed, so *obviously* speed, since speed is the ratio of elapsed distance to elapsed time—how far something has gone and how long it has taken to get there. Leaving her partner's hands, our ballerina shoots up; she then slows down on approaching the top of her leap, and after turning in the air, speeds up again until with an urgent exhalation she lands on her partner's toes. Her speed is changing over the course of her jump. But the definition of speed as the ratio of elapsed distance to elapsed time depicts her *average* speed. It does nothing more.

The calculus begins when a request is made for that ballerina's speed at a particular moment, her *instantaneous* speed, her speed *right now*. The request should, and often does, prompt an obscuring flurry, as when a dust-covered table is given a good solid whack. At any given moment—*that moment precisely*—the ballerina has not gone anywhere and she has not used up any time in getting there. The ratio of distance to time should therefore be 0/0. This expression is mathematically meaningless. This suggests that she is not moving at all, a declaration that we mathematicians must reject, if only from the sense that if she is not moving at any particular moment, how could she be landing with that dreadful thump?

In a section of the *Principia Mathematica* titled *The Nature of the First and Last Ratios,* Newton addressed this objection. His argument has all the force of his implacable genius. If an object in motion has no discernable speed at any given time, then "by the same argument," Newton observed, "it may be alleged that a body arriving at a certain place, and then stopping, has no ultimate velocity; because the velocity, before the object comes to the place, is not its ultimate velocity; when it has arrived, there is none." This, Newton remarks, is absurd.

And so it is.

In 1684, two years before Isaac Newton finally sent his *Principia* to the printers, Gottfried Leibniz published a paper titled "Nova Methodus pro Maxima et Minimus" ("New Methods for Finding Maxima and Minima") in the third volume of the *Acta Eruditorum*. Leibniz was at the time forty-two, and already a stately European intellectual. A prodigy in childhood, he had become a polymath as an adult, handling philosophy and mathematics and a dozen other disciplines with easy familiarity. Aspects of the calculus had been in the European air for more than fifty years, but like a series of scattered wisps, they had refused obstinately to cohere into a cloud. Leibniz had noticed those wisps in the early 1670s, his unpublished papers revealing a man of very great mathematical ability sensing in some frustration a pattern that he could not quite identify. In 1674, Leibniz recounts, a letter written by the mathematician Amos Dettonville (a pseudonym used by Blaise Pascal) prompted the wisps to gather into a thunderhead. He required time for reflection; in the "New Methods" the cloud at last burst.

The Leibniz of the "New Methods" was not concerned directly with distance or with speed; it is *curvature* that has commanded his attention. Time may for the moment be canceled. Let the ballerina remain panting in the wings. Now curvature is a measure of how radically a line is bending; and being bent is a property of curves that can very nicely be described by something so simple as the slope of a straight line, one cutting the curve twice and so forming a suspension bridge between points. Such are the secant lines of

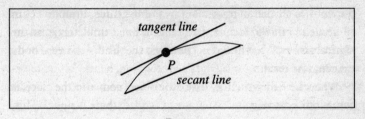

FIG. 4.1

plane geometry. Their slope measures change along one axis of a coordinate system juxtaposed to change along the other axis (Figure 4.1).

But the slope of a secant line may *also* be used to describe the average speed of a particle moving along the very same curve; it is, in fact, the very definition of average speed and so its reduction to number.

Speed—*squat, lift, grunt, up, down*—and curvature—*being bent*—are one and the same.

Do you see the connection? Well, you should.

And you should see as well that the enigma of speed is again the enigma of curvature. If the suspension bridge of a secant line makes for a rough-and-ready assessment of curvature between two points, it says nothing—how could it?—about curvature *at* a point, the way in which a sinuous shape deforms itself *right there.* Still, secant lines, inasmuch as they are straight, usefully suggest the importance of being straight in the scheme of things. The question is, of course, *which* straight line can by being straight convey something about being bent? The candidate with all the right connections is a straight line meeting the curve at just the point in question—*right there;* this is its tangent line.

Meeting a curve at a particular point, a tangent line shares its coordinates with the curve, the curve and the tangent by touching one another obliterating their separate selves. Leibniz had read Descartes; he knew that straight lines are governed by the equation

$y = mx + b$. He had in possession two of the three variables (x and y) needed to fix the identity of a tangent line completely in an analytical amber. What he did *not* have was the third—the *slope* of the tangent line itself.

When he had gotten it, that *boom!* boomed, and the calculus came into existence.

There is a place in mathematics where doubtful and deviant ideas are collected. I might as well call it Queer Theory. There are infinitesimals, for example. These are numbers smaller than any given number and yet not zero. If an infinitesimal number is added to itself, no matter how many times, it remains less than a given number. Writing more than two thousand years before the advent of the calculus, the Greek mathematician Archimedes had proscribed infinitesimals, his interdiction coming to be known as the Archimedean axiom. Although it is not all that easy to say why, a number smaller than *any* number and yet not zero offends common sense without quite managing to outrage it. Nonetheless, it is to Queer Theory and infinitesimals that Leibniz turned for an explication of being bent.

Now the slope of a straight line—and secant lines are surely straight—represents a ratio of finite distances. The idea of a ratio, Leibniz appropriated, but with this difference. The slope of a *tangent* line at a point, he argued, should be represented as the ratio of *infinitesimal* distances along the axes of a coordinate system. These ghostly distances he denoted by the symbols dx, dv, dy, where d indicates the business of shrinking a variable down to infinitesimal size, and x or v or y the variable being shrunk. The ratio of one infinitesimal to another, and so the requisite slope of a tangent line, is just dy/dx, a formula now famous in all of mathematics. "We have only to keep in mind," Leibniz wrote, "that to find a *tangent* means to draw a line that connects two points of the curve at *an infinitely small* distance."

Nous considerâmes, as Leibniz might have written to his snooty

French friends, the ample parabolic curve $y = x^2$. Down one side of the coordinate system it goes when x is negative, touching the origin at $x = 0$, and up the other side it goes when x is positive. At the point $x = 2$, $y = 4$, the parabola, having already changed elsewhere in the plane, is changing right there.

Ah, but *how*? How *precisely*?

With his luxuriant brunette wig cascading to his shoulders, and his fine face, the nose very large, alive with intelligence, Leibniz reminds us that he means to determine the slope of a line tangent to the curve at the point $x = 2$ and $y = 4$. His performance is half mad and half magic. And it *is* a performance.

A ratio has two parts, a bottom and a top.

The bottom. Start at 2 and go an infinitesimal distance dx along the x-axis. The infinitesimal change in x is $(2 + dx) - 2$.

The top. Square the bottom, part by part, getting $(2 + dx)^2 - 2^2$.

Take note. $(2 + dx)(2 + dx) = 4 + 4dx + dx^2$.

Allow a few dots . . . to accumulate.

Top to bottom. $(4dx + dx^2)/dx$.

Divide. $4 + dx$.

The magic. Drop dx from the last line because it is, after all, *infinitesimal*, leaving the number 4.

The mathematics. The number 4 is the correct answer, as it happens, the number corresponding to the slope of the tangent line, the curvature of the curve *right there*, its shape and speed reduced to number.

Applaud.

Whatever their logical status as things too small to be seen yet not too small to be counted, the invocation of infinitesimals allowed Leibniz to press his nose against a curve and by pressing gain a sense of its secrets at a specific point. His success was local, one act covering one point. But the "New Methods" is among other things the record of his attempt to convert a particular calculation into a general calculus, an almost mechanical scheme of reasoning. Absent such a scheme, Leibniz realized, calculations would tend to multiply; in Descartes, they had already become tedious. For a man who knew *his* calculus, Leibniz observed with satisfaction, "such calculations are easy to investigate."

If they were easy to investigate, those calculations, they were rather more difficult to explain. Writing in 1684, Leibniz lacked a concept that he would himself not bring into consciousness until 1695. It is the concept of a mathematical function. Together with the numbers themselves, functions are the most important of mathematical objects, and, as one might expect from the fact that Leibniz found it difficult to see them clearly and see them whole, they are intellectually elusive. The verbal gestures needed to explain things are easy enough. A function is a relationship, a scheme of coordination, a displacement of attention from one number to another, a rule, a regularity, or a plan. There is, for example, the twofold operation of *taking* a number and *squaring* it. The number one passes to itself, the number two goes to the number four, the number three to the number nine, and there is nothing at all preventing the energy that is responsible for these mental acts from being continued indefinitely. The specific conveyance achieved by this function mathematicians write as $f(1) = 1, f(2) = 4, f(3) = 9$, with the Roman letter f bearing down on a numeral and sending it to the numeral expressing its square. In addition to $f(1) = 1, f(2) = 4, f(3) = 9$, there is, as well, $f(x) = x^2$, a prescription for getting things squared away in general, the very expression in symbols of the very activity of squaring numbers.

The notation encompasses all of the elementary mathematical acts.

But neither my own verbal gestures, nor any table of examples, succeed in providing what is really needed, and that is a moment of complete intellectual clarity, circumstances that may be read backward into the late seventeenth century, as Leibniz used one concept that he could not precisely define to explore other concepts that he could not precisely see. Two centuries were to pass before Georg Cantor was to discover the words that in 1684 Leibniz lacked; during all that time, mathematicians continued to use functions of the most remarkable variety, indifferent to their own inability to capture the concept completely. What follows is thus an exercise in anachronism. I am explaining what Leibniz thought as he would have wished to have thought it; in this way, the calculations that he imagined were easy to explain turn out against every expectation to be easy to explain.

And now the famous formulas follow, formulas because they are little computational machines, and famous because they are still in use. The function $f(x) = x^2$, let us recall, ground to halt at $x = 2$, whereupon Leibniz exercised his hocus-pocus to determine that being bent *right there* could be described by the number four. This number described the slope of the line tangent to the curve at that very point. Leibniz now settled the question of curvature for *every* point on the parabolic curve, and not just one. Starting with the function $f(x) = x^2$ he determined that no matter the x, its curvature (or speed) at every point could be described by *another* function, and so he established a higher-order connection between functions themselves and not just between numbers. That other function is $g(x) = 2x$, and just as one might expect, at $x = 2$, $g(x)$ *is* four. This is no mere computational trick. More than fifty years before, Galileo had demonstrated that the distance covered by a falling object is proportional to the square of the time it has been falling. This I have already noted. It is now plain that a *function* is at work, one that for every instant down to the last instant of recorded time returns with a number measuring distance. That function is $f(t) = 16t^2$. The variable x has in this formula been retired, the variable t taking its place in order to signify typographically the underlying connection to the t of *time*. But having seen through to the identity between curvature and speed, we can now appropriate Galileo's formula ourselves and recover speed from *distance* by means of the function $g(t) = 32t$. It is this function that describes *speed at every single moment*. Like the original function measuring distance, this function, measuring speed, may itself be inscribed on the axis of a coordinate system, so that both distance and speed now have a visual identity, a place in the world of curves and so a place in the world of things.

Mathematical trifle though it may be, the operations just recorded were inaccessible to the entire human race until the latter half of the seventeenth century. They required the eye of genius in order to be seen. This part of the calculus—its famous formulas, computational tricks, and its staring prophetic eye—is called the differential calculus. The exchange between functions and their derivatives reduces itself to a list, one found on the inside cover of every calculus textbook, the place where Leibniz' noble genius is immured.

Differentiation involves the two ancient mathematical operations of subtraction and division, and a certain late-seventeenth-century sleight of hand. But subtraction and division are only two of the four great primitive operations of mathematics. There remain addition and multiplication. A sense of symmetry, if nothing else, might suggest that as differentiation goes forward from a function to its derivative, some other process, when conjoined with a fast shuffle of its own, might go back from a derivative to a given function. Many seventeenth-century mathematicians had seen or sensed that differentiation must have an inverse, an operation going smoothly back. They were, those mathematicians, unable completely to sharpen their insights into a theorem. Addressing the future, Descartes had asked mathematicians yet unborn to solve what he called the inverse tangent problem, and with a peevish sense of his own inadequacy, he had suggested that no mathematician would solve the problem in all the tides of time.

"Descartes," Leibniz observed tartly, was in the habit of "speaking with a little too much presumption about posterity."

As, indeed, he was, for having gone so far forward, Leibniz found the way back as well.

A Cartesian coordinate system is again in place. All dapper coordination, a function $f(t)$ is given, and after that the curve that it

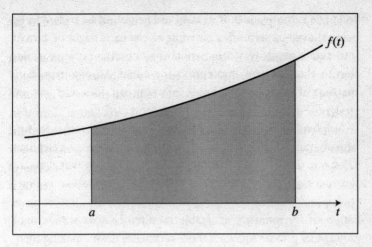

FIG. 4.2

expresses. The straight lines bounded by *a* and *b* form a canopy with the curve (Figure 4.2).

It is helpful to imagine that $f(t)$ depicts the speed of a moving object, but like all crutches, this one hinders as much as it helps, if only because it fastens the imagination on a particular example when what follows is wonderfully general, encompassing almost all continuous processes in nature. Keep that crutch if you must; discard it when you can; limp if need be.

Writing in the *Two Sciences*, and much occupied with problems in dynamics, Galileo argued that if $f(t)$ measures speed, it is the *area* underneath this curve that corresponds to the distance that an object has traveled. Area, after all, represents something like the product of a figure's base and height—*something* like it, *no?*—and the product of speed and time, what is it if not distance? A confused grope-and-fumble follows as Galileo struggled with a number of infernal infinitesimals of his own. He is going the right way but he is not on the right track.

The area underneath a curve, Leibniz argued fifty years later, *can* be approximated in antecedently familiar terms. He then proceeded to show how. The area of a perfectly ordinary Euclidean rectangle is the product of its base and height, and a rectangle, by being shoved underneath a curve, as an elephant might be shoved into a stall, stands very commendably as a first linear approximation to the true area underneath the curve. An *approximation,* meaning it is close; and *linear,* meaning its sides and top are straight (Figure 4.3).

Subdividing the rectangle into smaller rectangles makes the approximation better, just because less of that embarrassing elephant is left out. Once Leibniz had determined that cutting that elephant down to size must be a matter merely of reducing the width of various approximating rectangles, the way was clear for another invocation of *his* infinitely adaptable infinitesimals, as with a tight conceptual grip he squeezed those rectangles down, making them smaller and smaller until, at last, he had made their width infinitesimal. He then proposed to consider the sum of infinitely many rectangles whose height he was unable precisely to specify and whose bases he was unable to compute. Unvexed, Leibniz argued in favor of the existence of that sum and with a smile of contentment pronounced it *the* area underneath the curve.

An *area* and so a *sum,* a *sum* and so a *number.*

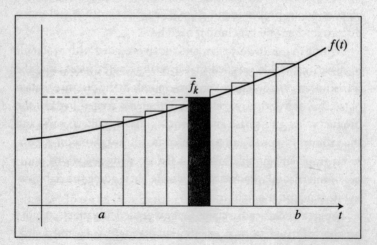

FIG. 4.3

The sum is known as the definite integral of a given function between the points *a* and *b*. Leibniz chose an elongated S to represent those sums of his, writing

$$\int_a^b f(t)dt$$

to signify the fact that an ordinary, old-fashioned exercise in multiplying the base *dt* by the height *f(t)* of a rectangle was now incarnated infinitesimally.

If the derivative squeezes change to the tip of a point, the definite integral assesses change over the expanse of an area. A flicker has given way to a pause.

But whether things are flickering or pausing, both the derivative at a point and the definite integral between two points return the mathematician to very particular numbers. Pythagoreans take note.

This part of the calculus is known as the integral calculus. It is technically more difficult than the differential calculus. If the idea of area is simple, and it is, the techniques needed actually to compute the area underneath a curve can be surprisingly rebarbative. Generations of students have loathed the definite integral on sight. Their disposition is not often improved when they encounter integration by substitution or integration by parts.

The area underneath a curve is a number and so something specific. It offers a frozen image of a fluid situation. Now the derivative is also engaged initially in a number-to-number transaction, one assessing a function's rate of change at a particular point. Thereafter that number-to-number transaction is displaced by one taking functions to functions. Motivated by the same desire to get out from under the particular, Leibniz—and Newton, too—is now vouchsafed one of those insights that changes the face of thought forever, at once revealing Euclidean geometry as a static discipline and breaking free of its hieratic time-bound constraints. If the area underneath a curve is changing at every moment, then it, too, should be represented by a *function*. That is just what functions do. They represent change. Now is the time for one of them to do it.

The definite integral, which until now has measured a fixed area, is thus enlarged so that it accommodates an area that is itself changing. This the mathematician indicates by writing the integral with a variable *t* where before there was only a fixed number. The result is the *indefinite* integral between a fixed point and a moving target:

$$\int_a^t f(t)dt.$$

A higher-level coordination of concepts is now about to take place, one that directly associates integration and differentiation and shows that they are inverse operations. The derivative of a function is a matter of subtraction and division; the integral, a matter of addition and multiplication. As no one at all might expect, these operations are linked at the deepest level. The second undoes what the first has done. This is a result that demands the beauty and perfection of mathematical notation:

$$\frac{d}{dt} \int_a^t f(t)dt = f(t).$$

This elegant, infinitely powerful symbolic statement affirms that two things are one. On the right, there is a function, one describing some continuous process. This is one thing. On the left, there is the derivative of its indefinite integral. This is the second thing. And when integration and differentiation have both been allowed to do their work, they turn out to be the same thing.

These ideas at once apply to the moving ballerina discarded some pages ago. There are three functions at work, the first expressing distance, the second, speed, and the third, area. There are two operations at work. The first is differentiation, the second, the expanding version of integration just noted. And now functions and operations are united so that the various fragments of this fresco, which until now have resisted assimilation, assume the aspect of a single figure. The area underneath the ever-changing curve measuring speed is given by the indefinite integral. But the indefinite integral of speed is distance, the derivative of distance is again speed, and the integral of speed is again distance. They are these things when they are understood as relationships, and so forms of change.

This is the fundamental theorem of the calculus.

Isaac Newton used the calculus in order to construct the *system of the world* in his *Principia Mathematica,* and within years of its discovery in the late seventeenth century, it was in wide use throughout mathematical physics. Newton introduced two laws of nature to the world. They are both laws of force. The first establishes the identity between force, on the one hand, and the product of mass and acceleration, on the other. The second establishes the universal law of gravitation. Objects in space, no matter how widely separated, attract one another with a force that is proportional to their mass and inversely proportional to the square of the distance between them. These equations lead at once to a system of differential equations, and differential equations are the chosen instruments of the physical sciences.

All drooping swan wings and tired feet, that ballerina may be allowed her final farewell. For suppose that having been lofted aloft, she is about to start down. At any moment of time how far will she have gone? It is a question that may be entirely enfolded in a mathematical method. She has, that ballerina, reached a certain height, one that we may symbolize by the number ξ, whatever the number, and whatever the units of measurement. The function $g(t)$, recall, denotes the rate at which she is changing her position in space at every moment, and so her instantaneous speed. What is desired is an unknown function x in which change in place has been directly correlated with change in time.

The wish is father to the symbol:

$$\frac{dx}{dt} = f(t),$$

an equation in which what is unknown in x is a function, a continuous coordination of events taking place in the real world.

The solution is immediate:

$$x = \xi + \int_a^t f(t)dt.$$

And so, too, the identity of x, which is simply the function $g(t) = 16t^2$.

And this is in accord with common sense. The distance covered by a falling ballerina is measured as the sum of her initial height, and the distance she covers thereafter, whatever the elapsed time, the coordination now perfectly general, infinitely flexible.

This is the instrument that has made possible Newtonian mechanics, Clerk Maxwell's theory of the electromagnetic field, Einstein's theory of general relativity, and quantum mechanics, the movement of the planets in the night sky and the atomic explosion over Trinity both encompassed by the same mathematical method.

Mathematicians and physicists contemplating the calculus and the theory of differential equations were uneasily aware that an instrument of great usefulness was somehow compromised by the fact that its chief concept made no sense; it was a point made with some force by the philosopher Bishop Berkeley, who scoffed at the calculus, calling infinitesimals the "ghosts of departed quantities." Like a true philosopher, Berkeley did not concern himself with the question why those ghosts should have been quite so lively. More than two hundred years were to pass before mathematicians could honestly say that they had provided a logical analysis of the calculus commensurate with the work that it was required to do.

That analysis having been initiated by Augustin Cauchy and then completed by Karl Weierstrass in the nineteenth century, infinitesimals disappeared in favor of limits. The requisite definition is very complex and it requires a good deal of reflection and practice before it can be assimilated, whereupon among mathematicians, at least, it is promptly forgotten. I shall now offer readers the chance to do some forgetting on their own.

The series of numbers $S = 1/n$ is plainly getting smaller and smaller as n is getting larger and larger. Destination zero. It is there that S reaches a limit L. Analytic understanding comes in three steps. First, take any positive number ε. Keep that number fixed for the moment. Then consider some other positive number δ, one that depends on ε, so that the two numbers form a team. The second step. Now consider the accordion formed between zero and $S = 1/n$ as n gets larger and larger. This is the third step. Benefits to follow.

The definition: L is a limit of S if there is for *any* choice of ε *some* choice of δ, such that for *all* values of n greater than δ, the accordion is less than ε.

Benefits proffered: With the definition of a limit, both the derivative and the integral of a function may be conveyed in purely arithmetic terms. Infinitesimals are gone. The derivative is instead depicted as the limit of a series of real and finite ratios, the ones formed by secant lines as they get closer and closer to a tangent line. The integral in turn is defined as the limit of a series of real and finite sums, the ones formed by those approximating rectangles. The delta-epsilon definition of a limit has become the foundation of a great many proofs, a subtle tool, but not one, even among mathematicians, that inspires affection.

Benefits withheld: The definition of a limit is difficult. And difficult in an odd way. It requires that four numbers be kept resident in memory (ε, δ, n, and L), and that the result of one mathematical operation be balanced against an inequality, even as two quantifiers are juggled in the background. The idea of a number greater than zero, but less than anything else and never mind what it means, is far more intuitive. Just why should such a complicated definition be required properly to explain what seems to be a simple idea?

Why indeed?

Working independently, but with the uneasy sense that some other grunting giant was lumbering off in the woods, Leibniz and Newton both outlined the major concepts of the calculus and stated and then proved its fundamental theorem. Their hands, like enormous hairy paws, are still all over mathematics. As the soul-shattering importance of their discoveries emerged in the late seventeenth and early eighteenth centuries, both men were naturally eager to appropriate full credit for their research and like candidates for the Nobel Prize, wasted little time before enlisting their sycophants in a campaign to darken the other's reputation. Newton was implacable, pestering his friends on the matter of priority, grumping loudly in his chambers, and generally carrying on in a fury. Leibniz' death in 1716, although it removed the source of his obsession, did nothing to dilute the intensity of his indignation. Leibniz was altogether more good-natured than Newton, and more comfortable in his skin. He was aware that in Newton he was confronting a formidable intellect and, were it not for the fact that he justifiably considered himself Newton's equal as well as his rival, he might have let things slide.

But really the issue is dead, and it should never have been alive. Both men were great.

5

COMPLEX NUMBERS

DID I HAPPEN TO mention that the Italian mathematician Girolamo Cardano was a splendid plagiarist? He is there in history, bright, capable, and industrious, but his eyes are fixed on another man's work. Mathematicians of the early sixteenth century knew how to solve ordinary algebraic equations. They thought in terms of minor mysteries, problems in identification in which a mathematical unknown leaves traces of itself in an equation, the mathematician noting that *this x,* whatever *it* is, seems to be none other than the number already in the drawing room, warming its hands before the fire. They used a variety of tricks. Like real detectives, they did a good deal of guessing. They knew as well how to solve quadratic equations of the form $ax^2 + bx + c = d$ by means of a little algorithm. The algorithm is now well known: $x = (-b \pm \sqrt{b^2 - 4ac})/2a$; and the device that it expresses works mechanically. Numbers go in on the right, solutions come out on the left. They were intrigued, those mathematicians, by cubic equations such as $x^3 + mx = n$ in which amid the two parameters m and n an unknown x finds itself taken to a third power.

In the early part of the sixteenth century, Scipione del Ferro, a professor of mathematics at the University of Bologna, let slip hints about a fabulous formula that he had discovered, something that could solve cubic equations. Matters were kept closely guarded. Ferro passed his secret to his pupil, Antonio Maria Fior, who proceeded to demonstrate that the chief motive for betraying a secret is the desire to boast of having been vouchsafed a confidence. He boasted shamelessly. In 1535, another mathematician, Niccolò Tartaglia, worked out Ferro's formula for himself. He kept the result close to his vest. A very competent mathematician in his own right, Cardano was consumed with curiosity: he pestered Tartaglia endlessly and on Tartaglia's telling, at least, promised that if he could acquire the formula without effort, he would guard it without compromise. "I swear to you by God's Holy Gospels," Cardano solemnly affirmed, "and as a true man of honor, not only never to publish your discoveries, if you teach me them, but I also promise you, and I pledge my faith as a true Christian, to note them down in code, so that after my death no one will be able to understand them."

A year later, Cardano published the formula under his own name in a work he entitled the *Ars Magna*.

Mathematicians of the sixteenth century could make no obvious sense of the square root of minus one. No number in their experience when squared was less than zero. The methods of classical antiquity were unavailing. Yet the equation $x^2 = -1$ seemed stubbornly to suggest that beyond its five simple symbols, something was there. Other equations arising in the most natural way also called for the square roots of negative numbers. They resembled a series of startled house guests pointing to the same ghost by the very same high window. In setting himself the problem of dividing ten into two parts whose product is forty, Cardano began with the equations $x + y = 10$ and $xy = 40$. From these he derived the equation $x(10 - x) = 40$. The roots of this equation are $5 + \sqrt{-15}$ and $5 - \sqrt{-15}$. The quadratic formula suffices. Negative numbers again appear under a radical sign. "Put aside the mental tortures involved," Cardano advised himself, and manipulate these expressions as if they made sense. In some way, he reached the correct conclusion. "So progresses arithmetic subtlety the end of which, as is said, is as refined as it is useless."

The same unnerving combination of square roots and negative numbers appeared when Cardano put his stolen formula to use, punishment for theft, I suppose, and an unmistakable indication that by means of this formula nature was endeavoring to call attention to itself. Although complicated, Tartaglia's fabulous formula specifies only very simple mathematical operations. Fractions are at work, roots extracted, powers raised. For any equation of the form $x^3 + mx = n$,

$$x = (n/2 + \sqrt{(n/2)^2 + (m/3)^3})^{1/3}$$
$$- (-n/2 + \sqrt{(n/2)^2 + (m/3)^3})^{1/3}.$$

The formula at once leads to the same dark wood. Consider the equation $x^3 - 15x = 4$. A straightforward computation shows that

$$x = (2 + \sqrt{-121})^{1/3} - (-2 + \sqrt{-121})^{1/3},$$

with negative numbers again appearing underneath radical signs. Cardano knew they were there; he did not know what they signified.

His contemporary, Rafael Bombelli, discussed the same equation and came achingly close to seeing clearly the problems of the eighteenth century while living in the sixteenth. He knew by guessing that $x = 4$. And with four in hand, he could work backward *from* $x = (2 + \sqrt{-121})^{1/3} - (-2 + \sqrt{-121})^{1/3}$—the solution specified by the formula—*to* $x = (2 + \sqrt{-1}) - (-2 + \sqrt{-1})$. When negative and positive signs interact in this expression, square roots vanish by mutual annihilation. For a moment, the mathematical landscape that to his contemporaries remained inky and obsidian must have seemed suffused with light as those complicated symbols resolved themselves into an ordinary number. Bombelli had used the solution of a cubic equation to obtain the solution of the same equation by means of Tartaglia's formula. He had seen the machinery of the formula as if it had been shielded by a transparent pane; but without the solution, he was unable to set it in motion.

Bombelli's masterpiece, by the way, remained inaccessible for more than four hundred years, appearing finally in French in 1929 under the title *L'algébra*.

During the seventeenth century, complex numbers, like the ghosts they were, appeared and disappeared, European mathematicians seeing them by dark stairs, or hearing them scratch behind the wainscoting, and then not hearing or seeing them at all. A number of notable mathematicians—Leibniz and John Bernoulli, for example—engaged in polemics with one another, trying in vain to determine whether the complex numbers whose existence they doubted had logarithms or whether the square root of minus one had a square root of its own. They both reached incorrect conclusions and championed them resolutely. They were unable to shake the odd feeling that they were debating the finer points of nothing.

And yet within a period of fifty years, those complex ghosts gave up their secret hiding places and with a presence unmistakably *real*, swept away all metaphysical doubts. The first step in their rehabilitation took place when European mathematicians learned just how complex numbers should be *represented*, thus verifying the useful philosophical principle that orthography recapitulates ontology. Complex numbers, mathematicians came to understand, take the form $a + bi$. A later addition to mathematical notation, the letter i denotes the square root of minus one so that $i^2 = -1$; the letters a and b designate ordinary numbers. The number bi is the product of the square root of minus one and b; and $a + bi$ is the sum of that product and a. If a is zero, $a + bi$ is purely imaginary; if b is zero, then completely real. The complex conjugate of a complex number $a + bi$ is simply $a - bi$.

The second step in that program of metaphysical rehabilitation now followed. Addition, subtraction, multiplication, and division are the most primitive and the most important operations in mathematics. Let us say that they are algebraic. Any number derived from complex numbers by algebraic operations, Jean Le Rond d'Alembert affirmed in a paper titled "Réflexions sur la cause général des vents," is again a complex number. Starting from complex numbers, one gets complex numbers. The circle is never broken by the intrusion of anything weird.

Addition and subtraction are defined among the complex numbers in an intuitive way:

Addition: $(a + bi) + (c + di) = (a + c) + i(b + d)$;

Subtraction: $(a + bi) - (c + di) = \ldots$,

the crutch of three dots covering the transmogrification of a plus to a minus sign and nothing more.

Multiplication follows:

Multiplication: $(a + bi)(c + di) = (ac - bd) + i(ad + bc)$,

with division three dots behind, although, perhaps, these three dots are more complicated than most.

Complex arithmetic—what else to call it?—may well seem as if a series of utterly arbitrary rules were being imposed on an object of almost impudent insubstantiality, an effort rather like drawing a pedigree for unicorns. This impression is mistaken: The rules are *not* arbitrary. The addition of complex numbers is *ordinary* addition, the rules governing $(a + bi) + (c + di)$ governing $[3 + (7 \times 8)]$ $+ [5 + (2 \times 8)]$ as well. So, too, the other arithmetic operations.

The complex numbers, as mathematicians say, are closed under the usual algebraic operations. The date is 1747. Unlike poor Bombelli, d'Alembert had the enormous success of seventeenth-century mathematics acting as wind to his sails. Still neither d'Alembert nor anyone else advanced any reasons whatsoever for supposing that behind the symbol $a + bi$ lay more life than might be found in the printer's ink with which it was written.

The unicorn is still a unicorn, but it helps to see that its pedigree is impeccable.

The complex numbers may be added, subtracted, multiplied, and divided. There remained the transcendental operations. What are we to say, for example, when a mathematician takes it into his head to ask for the meaning of $(5 + \sqrt{-5})^{\pi i}$? A real number taken to a real number is simply the real number multiplied by itself, so that 2^3 is just $2 \times 2 \times 2$. It seems hardly profitable to suggest that $(5 + \sqrt{-5})^{\pi i}$ is just $(5 + \sqrt{-5})$ multiplied by itself πi times. "In this way," as Lipman Bers once gravely remarked in class, "lies madness." What, for that matter, is the logarithm of a complex number, or its sine or cosine? These questions had been long answered for the real numbers. No one expected, when they were posed anew for the complex numbers, that they would lead to a dazzling unification of experience. It was nonetheless so.

It is now roughly the middle of the eighteenth century, the era that belongs to Leonhard Euler. The largest mathematical personality of his time, Euler was born in 1707 and died in 1783. He spent the greater portion of his active life at the St. Petersburg Academy, where he had initially been invited in 1727 by Catherine I; but at her unexpected death, the Russian government passed into the hands of men who regarded a scientific academy with suspicion, and Euler, fearful of the spies and gossips at court, was forced to live in social isolation, work his joy and marriage his refuge. In 1741, Euler accepted an invitation from Frederick the Great to join the imperial court in Berlin; he retained a living memory of his dark Russian days, remarking once to the dowager empress, when she questioned him on his unforthcoming silence, that, "Madame, I come from a country where, if you speak, you are hanged." When Catherine the Great ascended the Russian throne, she invited Euler to return to St. Petersburg, and when he did so gratefully, she treated him as visiting royalty, which, of course, he was. It is there that Euler passed his final years. He remains living in thought in terms of the splendid example he provided of a mind capable of movement without friction and so achievement without effort.

Euler played brilliantly with everything in mathematics, but he played most brilliantly with complex numbers, showing that far from comprising mathematical oddities, they were instruments that providence had provided for the recovery of lost symmetries, unsuspected connections, a sense of the unity of experience. This is most evident in the relationship he discerned between the trigonometric and exponential functions.

In elementary textbooks, trigonometry is given over to a number of formulas dealing with the properties of a right triangle. In this context, they are largely incomprehensible, if only because they are largely unmotivated. Those texts, and the memories that they convey, must be canceled and abjured. The trigonometric functions serve manfully to ferry real numbers to real numbers. They are in the business of relationships. It is useful in this regard to get rid of degrees and think of angles in terms of radians. A circle of 360 degrees has 2π radians; a half circle of 180 degrees, π radians, and a right angle, $\pi/2$ radians. The radian, as a unit of measurement, may also be ejected from this discussion, leaving behind only the pure real numbers. Sine, cosine, and tangent now operate cleanly as functions. The sine of π is zero, and the cosine of π is minus one, and so similarly for all other real numbers. These functions are periodic, moving between plus and minus one, and repeating their pattern endlessly (Figure 5.1).

The exponential functions are quite different. A real-valued function $f(x)$ is exponential if it has the form a^x. The exponent x is up there in the cockpit; the exponee a down there in the baggage compartment. Since the parameter a may be varied, the exponential function is really a family of functions, a whole clan. When it comes to exponentiation, multiplication may always be expressed in terms of addition, so that $(a^x)(a^y) = a^{x+y}$. This is a clue as to what is coming and, like all the best clues, it is there in plain sight. Plainly, the exponential functions are *not* periodic. They mount up inexorably, one reason that they are often used to represent doubling processes in biology, as when undergraduates divide uncontrollably within a Petri dish (Figure 5.2).

Curiously enough, the various exponential functions may be amalgamated into a single function—*the* exponential function. The connecting tissue is formed from a single number. Like the number π, the number e is just one of those things. It is a real num-

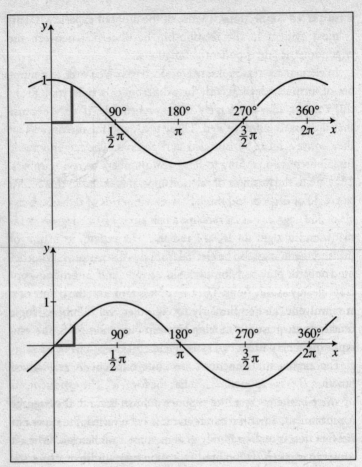

FIG. 5.1

ber, and so ordinary, but like the square root of two, it is irrational and its decimal expansion, 2.71828 . . . , goes on and on, as formless as the wind. It appears everywhere in mathematics, where quite literally it forces itself on the imagination, and like π, *i*, 0, and 1, it seems to play a role similar somehow to a fundamental constant in theoretical physics.

Through the magic of mathematical definition, *any* function a^x can be expressed entirely in terms of the number *e*. Only two steps

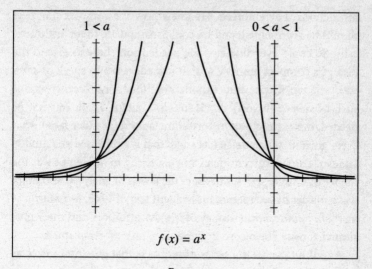

$$f(x) = a^x$$

FIG. 5.2

are involved. The number a is first expressed as $e^{\log a}$, where $\log a$ is taken to the base e. And then a^x is expressed as $e^{x \log a}$. The various exponential functions have just disappeared in favor of a single exponential function, $exp(x)$ or e^x—the number e raised to the power x.

Whatever the nature of the trigonometric and exponential functions, "the person who approaches calculus from the point of view of real numbers," the contemporary mathematician Lars Ahlfors has observed, "will not expect *any* relationship between the exponential function and the trigonometric functions." Indeed, he adds at once, "these functions seem to be derived from completely different sources and with different purposes in mind."

That this impression is mistaken represents a remarkably thrilling discovery in the history of thought.

By the fourth decade of the eighteenth century, Euler had at his fingertips the wealth of European mathematics, its superb achievements, its complicated (and often incoherent) definitions, and the distinctions they enforced. He knew how the complex numbers should be represented; and he could manipulate them arithmetically. So could everyone else. He needed somehow to extend the idea of a complex number so that it made sense to speak of *complex* trigonometric and exponential functions—constructions such as **sine** $(5 + \sqrt{-5})$ or $(5 + \sqrt{-5})^{\pi i}$, and if *I* talk of what *he* needed, this is purely an historical metaphor, for Euler faced what every great mathematician faces, and that is a confused and muddy tableau. The instrument that came naturally to his genius was the theory of infinite series. These are the jewels of analysis. And Euler was a master of such series, his brilliant text of 1748, *Introductio in analysin infinitorum,* a display of insight, intuition, and sheer musical virtuosity, the record in symbols of an exuberant spirit.

An infinite series is a series of numbers that goes on forever, as when $S_n = 1, 1/2, 1/3, \cdots, 1/n$, the mathematician's by now familiar three dots abbreviating a process that although without end is *not* without meaning. S is the name of the series and the subscript *n* indicates that terms in the series are formed by allowing *n* first to be one, then two, then three, and so on to the inevitable *and so on.* The series S_n, although infinite, has an easily recognized form. It is tending somewhere. As *n* gets larger and larger, the terms of the series get smaller and smaller, the series tending at the end of time toward the number zero.

These very plausible ideas may now be used to give life, and so meaning, to the concept of an infinite sum, an object of the form $S_n = a_1 + a_2 + \cdots + a_n + \dots$.

Unlike the infinite series $S_n = 1, 1/2, 1/3, \cdots, 1/n$, the infinite sum $S_n = 1/2 + 1/4 + 1/8 + 1/2^n \dots$ demands a refined and elaborate interpretation, a way of coaxing a finite number from an infinitely continued process, and so it requires a definition. It is a definition that proceeds in stages, but the mathematician's art involves nothing more than a recovery of an infinite series from an infinite sum, so that what has in the end been accomplished is, like the emergence of a butterfly from a chrysalis, the unexpected consequence of something that has been there all along. The definition, like that butterfly, emerges in stages:

The first:
From $S_n = 1/2 + 1/4 + 1/8 \ldots$ form an associated series of partial sums:

$$1/2 = a_1,$$

$$1/2 + 1/4 = a_2,$$

$$1/2 + 1/4 + 1/8 = a_3.$$

The second:
Consider $S_n = a_1, a_2, a_3, \cdots, a_n$ as an infinite series.

The third:
Ask whether there is a limit toward which $S_n = a_1, a_2, a_3, \cdots, a_n$ is tending.

The last:
Assign this limit to $S_n = 1/2 + 1/4 + 1/8 \ldots$ as its sum.

These stages reveal almost by examination that the infinite sum of S_n is 1, a wonderfully concrete, palpably concrete number, the butterfly of the infinite achieving after pupation iridescent but finite wings.

This conclusion is in accord both with the definition just offered and with intuition, which, in examining the sequence, tends to loiter about the number one, even if the loitering is prompted by an impression that only the definition itself can articulate and make conscious.

Mathematical ideas now reveal an urgent inner logic, forcing the mathematician simultaneously to attend to a number of different voices, as in polyphonal music. The idea of a function is crucial, and it is crucial everywhere in mathematics. Sending numbers to numbers, a function designates the most primitive (and thus the deepest) form of mathematical life. An infinite sum has the power to express a function, and so bring about a coordination between numbers. This is by no means obvious and the development of the theory of infinite sums represents one of those decisive episodes in which human self-consciousness enlarged itself dramatically. Isaac Newton exploited the power of infinite summation in much of his mathematical work; and it is one of the oddities of mathematical history that while he understood that infinite sums define functions, he could not quite place his finger on the very concept of a function itself. In the eighteenth century, every mathematician's finger got to the right place, Euler's most notably.

The infinite sum $S_n = 1 + x/1! + x^2/2! + x^3/3! + \ldots$ contains in x an unknown number, and in this respect makes contact with the great clan of mathematical expressions in which something is missing and so something must be determined. In all other respects it is like the infinite sums already scouted and I am introducing the sum explicitly because it is destined to do great things. The exclamation point after numbers in the denominator, by the way, designates their factorial—the product of that number and all smaller whole numbers. Thus 3! is $3 \times 2 \times 1$. Everything else in S_n is ordinary. And yet, this infinite sum plainly expresses a relationship, one forged between x, whatever *it* happens to be, and the sum of the infinite series itself, whatever *it* happens to be. As the value of x changes, the sum changes as well. And is this not the essence of what is means to express a function?

This particular infinite sum happens to express the exponential function, so that

$$e(x) = 1 + x/1! + x^2/2! + x^3/3! + \ldots.$$

On letting $x = 1$, the mathematician expects to see e^1 in return; but any number raised by one is simply the number itself, as when 10^1 is just ten. At $e(1)$, the expectant mathematician looks for

2.71828 . . . , and this, it is gratifying to observe, is just what in the end he gets.

If infinite sums may be used to express the exponential function, they may also be used to express the sine and cosine functions. Different constructions are at work but just the same principle:

$$\text{Sine } x = x - x^3/3! + x^5/5! + x^7/x! + \ldots .$$

$$\text{Cosine } x = 1 - x^2/2! + x^4/4! - x^6/x! + \ldots .$$

These relationships may also be checked by checking sine and cosine functions for particular numbers—1, say, or 0, or π.

In all this, the seventeenth century has had its due. The extension of the exponential, sine, and cosine functions to the world of complex numbers is properly the work of the eighteenth century. The very first step is the one that is obvious. In the definition of the exponential and trigonometric functions by infinite sums, the variable x has always been a real number. It is now replaced by a complex number $z = a + bi$. The substitution is purely mechanical, one symbol standing in for another in the original definition of the exponential function by means of an infinite sum; it involves no thought:

$$e^z = 1 + z/1! + z^2/2! + z^3/3! + \ldots .$$

Nonetheless, a threshold in both history and thought has been reached.

Rafael Bombelli must be allowed to return from the sixteenth century for one last time. Although he never quite grasped the glittering gold he had in hand, Bombelli knew something about the square root of minus one; and in particular he knew that $i^2 = -1$, $i^3 = -i$, $i^4 = 1$, $i^5 = i$, . . . and so on up the ladder of increasing exponents. How he knew any of this, I have no idea.

To this fact must be added the clue already given and left lying in wait. The expression e^z is simply a convenient way of writing

e^{x+yi}. And e^{x+yi} is simply another way of writing $(e^x)(e^{yi})$. From these facts and that clue, it follows that

$$e^z = e^x e^{iy} = e^x [1 + iy + (iy)^2/2! + (iy)^3/3! + (iy)^4/4! + \ldots].$$

A dazzling connection between the exponential and the trigonometric functions now emerges into the colored light, for

$$e^{iy} = 1 + iy - y^2/2! - iy^3/3! + y^4/4! + \ldots.$$

Or what comes to the same thing

$$e^{iy} = [1 - y^2/2! + y^4/4! - y^6/6! + \ldots] + i[y - y^3/3! + y^5/5! - \ldots].$$

But the infinite sums within the first and second brackets represent nothing less than the infinite sums designating the cosine and sine of a function.

It follows that

$$e^{iy} = \cos y + i \sin y$$

so that

$$e^z = e^x(\cos y + i \sin y).$$

This is Euler's formula.

What a wealth of insight Euler's formula reveals and what delicacy and precision of reasoning it exhibits. It provides a definition of complex exponentiation: It *is* a definition of complex exponentiation, but the definition proceeds in the most natural way, like a trained singer's breath. It closes the complex circle once again by guaranteeing that in taking complex numbers to complex powers the mathematician always returns with complex numbers. It justifies the method of infinite series and sums. And it exposes that profound and unsuspected connection between exponential and trigonometric functions; with Euler's formula the very distinction between trigonometric and exponential functions acquires the shimmer of a desert illusion.

On a still deeper level of insight and analysis, Euler's formula leads to an even more mysterious relationship. If $x = 0$ and $y = \pi$,

$$e^{i\pi} = (\cos \pi + i \text{ sine } \pi).$$

But the sine of π is zero and the cosine of π is minus one; it follows that

$$e^{i\pi} = -1$$

whence

$$e^{i\pi} + 1 = 0.$$

This is the most famous formula in mathematics, linking in one simple statement the five most fundamental numbers and the basic concepts of addition, exponentiation, multiplication, and identity. It cannot be improved and words are inadequate to convey its beauty.

Nor are they necessary.

European mathematicians had learned to add, subtract, multiply, and divide the complex numbers, but even the greatest of the great mathematicians remained half-persuaded that in the square root of minus one, nature was exhibiting herself in a subtle disguise. "The true metaphysics of $\sqrt{-1}$," Gauss wrote in 1825, and so almost one hundred years after d'Alembert, "remains elusive." Fearful of some form of intellectual contagion, lesser mathematicians often encouraged their students to avoid such numbers altogether. It is at the beginning of the nineteenth century in a moment of wavering impulses that the logical development of complex numbers is enriched by the realization that these ghostly things could themselves be *seen,* a single diagram erasing entirely the impression among mathematicians that no matter how they might manipulate these numbers, they were not really there.

As so often happens in mathematics, the complex numbers were first seen in two different countries and by two different men, striking evidence that in mathematics the maturation of ideas is as much a social as an individual process. Both the Norwegian, Caspar Wessel, and the Swiss, Jean Robert Argand, were self-taught, intrepid amateurs, Wessel a land surveyor and Argand a bookkeeper. With their contributions made, they disappeared from the history of thought, leaving behind only their names in countless texts.

Wessel and Argand had roughly the same idea: The complex numbers, whatever they might be, could be precisely coordinated with points in the plane. Argand published his thoughts in 1806 in a monograph entitled *Essai sur une manière de représenter les quantités imaginaires dans les constructions géométriques*. A Cartesian coordinate system is again required. The origin of the system is zero, and real numbers crawl along both sides of the x-axis and up and down the y-axis. *Real* numbers—nothing more. Nonetheless, a number of the form $a + bi$ may now be given an instantaneous pictorial identity, a geometrical stamp, by virtue of the fact that the real number a is interpreted along the x-axis, and the real number b, along the y-axis. The complex number $a + bi$ thus corresponds uniquely to the point $\langle a, b \rangle$. If b is itself zero, so is bi, and in that case, the point $\langle a, 0 \rangle$ collapses into the real number a. Since the square root of minus one is entirely imaginary, it has the form $0 + 1i$. It corresponds to i itself. Starting from the origin, the mathematician need only go up one unit to reach the point $\langle 0, 1 \rangle$. It is right there in plain sight. No more than a dot, it has within its arithmetical identity all the power of something long sought but never found.

What I have sketched is only a bare beginning, one that corresponds to the initial moment in the development of a theory. The logical place in which these bare beginnings come to an end is the fundamental theorem of algebra. It is the fundamental theorem that asserts that any polynomial function $P(z) = a_0 + a_1 z + \cdots + a_n z^n$ has at least one root—one real or complex number, that is, making it true. The theorem's import is plainer than its symbolic expression. No matter how real and complex parameters and variables are combined, the theorem affirms, there is never a need to go beyond the complex numbers. No further numbers are needed. They suffice and so they bring to completion the very long effort at construction that over thousands of years yielded first the natural numbers, then the fractions, then zero and the negative numbers, and after that the real numbers. The complex numbers complete the arch.

Beyond the theory of complex numbers, there is the much greater and grander theory of the *functions* of a complex variable, as when the complex plane is mapped to the complex plane, complex numbers linking themselves to other complex numbers. It is here that complex differentiation and integration are defined. Every mathematician in his education studies this theory and surrenders to it completely. The experience is like first love.

I once mentioned the beauty of complex analysis to my great friend, the mathematician M. P. Schützenberger. We were riding in a decrepit taxi, bouncing over the streets of Paris.

"Perhaps too beautiful," he said at last.

When I mentioned Schützenberger's remarks to René Thom, he shrugged his peasant shoulders sympathetically.

This is one of the charms of the theory of complex numbers and their functions. It has broken men's hearts.

6

GROUPS

SOMETIME AFTER DAWN ON May 30, 1832, a young man named Évariste Galois, a pistol in hand, crossed a muddy field to the south of Paris and, after seconds had solemnized the arrangements, exchanged shots at twenty-five paces with the man who had asked him for satisfaction, Pescheux d'Herbinville. Galois had no very great abilities as a marksman. Plump and oily, d'Herbinville was a crack shot. His bullet pierced Galois' intestines. Galois fell to the ground, where he lay unattended: No one had thought to bring a surgeon to the duel. Spectators and seconds drifted off and, immensely satisfied with himself, so did d'Herbinville. Some three hours later, a passing peasant found Galois and took him by cart to *l'hôpital cochin* in the south of Paris. Dingy now, it was dingy then. Completely conscious, Galois waited his inevitable end from peritonitis. When his younger brother arrived and dissolved in tears, he said, "Don't cry. I need all my courage to die at twenty." And the next morning, he was dead.

Évariste Galois was a supremely original mathematician, and the story of his death has passed completely into myth, with only his death in some absurd duel proof against scholarly skepticism. Teachers had early observed in him signs of "mathematical madness," an uncontrollable passion akin to the divine madness of the Pythagoreans. He had twice been refused admission to the École Polytechnique, one of the influential schools established by Napoléon, because his examiners, although aware of his remarkable power, were unable to follow his thoughts. Galois did nothing to help his own cause. He was an indifferent speaker. He stood there sullenly in the examination hall and, when asked questions, fumed with barely suppressed fury, finally hurling a blackboard eraser at the unoffending head of one of his examiners. Bad luck had, in fact, been his companion, following him from early adolescence like a dismal black spaniel. Although a gifted classical scholar, every manner of authority that he encountered in his adolescence—professional, clerical, political—vexed and oppressed him. As a child, he had heard the drumbeat of Napoléon's cavalry thundering across European history, and through his father he had retained a vivid impression of the French Revolution. His generous imagination had been inflamed by Republican sentiments. During the France of his young manhood two men of uncommon mediocrity, Louis XVII and Charles X, had been returned to the French throne. They presided over a grim, colorless, pinched, mean autocracy. Galois found the narrowness and oppression of French life unendurable, and he saw in politics the same structures of pointless authority that he had found in academic life. He associated with cynical men and shadowy revolutionaries.

Although often described as "dangerous" by the police, he seems to have been innocent as the rain.

At the age of twenty, Galois lost his virginity along with his heart to a woman of reputed beauty but uncertain reputation, Stéphanie Félice Poterine du Motel. She was at the center of a number of obscure political factions. It was for the sake of her honor that d'Herbinville challenged Galois to a duel. The challenge was in effect a death sentence. This is something Galois knew. On the night of May 29, 1832, he sat at his desk and proposed to commit to posterity the teeming and obsessive mathematical ideas that he had until then kept locked within his skull.

He scribbled for hours, covering sheet after sheet with his slanting handwriting. "I have not enough time," he wrote in the margins; in the center of one page he wrote the words "*une femme.*"

Most men and women, if they remember high-school algebra at all, remember only a series of frustrating word problems, typically involving postmen or potholes, and a number of rules that seem often to lose themselves in a wilderness of symbols. A minus times a minus is a plus. *Yes, of course.* But why? It is almost as if some form of bizarre electromathematical repulsion were at work, the negative signs meeting in mutual annihilation. There *is* something in the world of algebraic manipulation that suggests obscurely fundamental exchanges of the sort that quantum physicists believe lie at the very heart of nature, one reason that algebra has acquired a supremely important role in mathematical physics. Although algebra is almost always taught to students too young by far to gain an enlightened sense of its nature, algebra is a profound mathematical discipline and by a trail as visible as a series of bright runway lights it leads directly to the world beyond mathematics. It is at least a part of modern algebra that Galois created in the final night of his life. Had he never lived or worked or loved, mathematicians would have discovered what he discovered, but they would have lost an idol, a young man as interesting as the young Byron, and as talented, and as doomed.

The problem that occupied Galois was not new. Mathematicians of the great ninth-century Moslem Renaissance knew how to solve a great many algebraic equations. They understood that all of mathematics may be reflected from within a single imperative, and that is to find an unknown (an injunction known to play a role in literature as well as life). It is for this reason that mathematicians place equations at the very center of their concerns, for an equation manages in a handful of symbols to express the whole of the mathematical drama itself. There is something. It answers to certain conditions. But what *is* it? Even the simplest of equations, such as $5x = 25$, expresses this inherent tension between what one has (five symbols) and what one seeks (that unknown something); and when such equations are solved, the exercise displays the same pattern of tension and release that is characteristic of all biological activity.

Dealing with a much wider budget of equations than the clever Arabs, sixteenth-century Italian mathematicians still thought in terms of minor mysteries, problems in identification in which a mathematical unknown leaves traces of itself in an equation. They used a variety of tricks. They did a good deal of guessing. And they knew how to solve equations of the form $ax^2 + bx + c = d$ by means of the quadratic formula.

They were perhaps unaware of the fact that for all its simplicity, the quadratic formula is a remarkable cage of symbols, indicating as it does that the solutions to a quadratic equation may be computed by manipulating its coefficients and *only* its coefficients. The familiar operations of addition, subtraction, multiplication, division, and root extraction are at work. With the coefficients in plain sight, these simple operations fix the equation's solutions; they determine its roots; they control its nature.

Solvability by radicals—that is how the technique came to be known, and the description has an eerie aptness, suggesting, as it does, a vigorous, a *radical*, effort to cut through those loitering coefficients and get to the heart of things directly.

There are, of course, equations galore within mathematics, and by the early years of the nineteenth century, mathematicians had gained the self-confidence to go beyond the ordinary. Quadratic, cubic, and even quartic equations, in which a variable x mounts itself four times? Over and done with. Mathematicians had formulas in hand. Cardano, the splendid plagiarist? A member of the Academy, his farsighted eyes notwithstanding. In the *Disquisitiones arithmeticae*, the great Gauss had studied with striking success equations of the form $x^p - 1 = 0$, where p is a prime number. Such are the cyclotomic equations, special cases of the more general Abelian equation $x^n - 1 = 0$, named in honor of the Norwegian mathematician N. H. Abel. The little streams and rivulets run through the sand, double back on themselves, freshen, and dry out.

There now occurs a curious division in the history of thought, something that mathematics can illuminate but never completely explain. Equations in which an unknown is taken to the *fifth* power, such as $x^5 - x^2 + 24 = 0$, baffled and confused Renaissance mathematicians. And they continued to baffle and confuse mathematicians in the two hundred years thereafter. In some mysterious way, mathematicians understood instinctively, nature draws a distinction between equations in which an unknown is raised to the fourth power and equations in which it is raised to the fifth. There is no question but that the roots of such equations exist. This Gauss had demonstrated by proving the fundamental theorem of algebra in four different ways, and so virtually pounding the poor thing to death. But could every such equation be *solved* by the method of radicals? This is entirely another question.

That night, Galois demonstrated that, *no,* quintic equations could not *necessarily* be solved by manipulating their coefficients arithmetically and fishing endlessly for their roots. In this he was duplicating work already undertaken by Abel and Ruffini. But *what* Galois demonstrated and *how* he demonstrated it are two entirely different things, for in order to dispose of an old problem, Galois was moved to the creation of a new idea. It is one of the profound ideas in modern mathematics, playing a role in all mathematical thought comparable to the role played in chemistry by the discovery of the molecular structure of matter.

It is the idea of a group.

The *idea* is as accessible and straightforward as the human nose. A group is a collection of objects, one that is alive in the sense that some underlying principle of productivity is at work engendering new members from old. The family is the primordial group beyond mathematics, and the divine domestic undertaking in which men and women reproduce themselves is the deepest, most primitive, and most mysterious operation in all of nature.

The sunshine-in-Venice atmosphere of these metaphors must now give way to the sleet-in-Scranton aspect of a complicated mathematical definition. The integers comprise the positive and negative whole numbers and zero, and no matter the integers, their sum is always an integer in turn. It is the integers *and* the operation of addition that taken together comprise a group and so form a single object of contemplation. A difficult mental operation is required to get that contemplation going. Practice is required, as in learning two-part harmonies on the piano.

Patience, too.

And a certain willingness initially to be defeated by intellectual experiences.

The formal, the mathematical, definition of the group comprising the integers under addition is grounded in one large fact and three subsidiary details:

The large fact:

Any two numbers may be added to each other, *yielding another number* in turn.

The first detail:

Addition is associative so that the order in which it is performed is unimportant: $(2 + 3) + 5 = 2 + (3 + 5)$.

The second detail:

Every number has an identity in zero so that adding zero to any number confirms the identity of every number. Thus $6 + 0 = 6$ and so, too, *any* number plus zero is just that number.

And the third:

Every number has a negative inverse so that adding a negative number to the same positive number results in mutual annihilation, with $-5 + 5$ leaving only an opalescent residue in zero.

That large fact and those three details describe a particular mathematical structure, the group **G** of integers under addition, the letter calling to mind the group, and the group calling to mind the concept. Now **G** has made an appearance against the screen of abstract thought hopelessly bound to a very particular, and so a very partial, mathematical object—the integers. There is no reason that the concept of a group cannot be peeled away from this example, and with the peeling away done, what emerges is the perfectly general and perfectly abstract idea of a group itself.

That general idea is only a few definitional steps away.

In place of those particular integers, there is an anonymous collective $G = \{a, b, c, \ldots\}$. Members of the collective are $a, b, c, \ldots$, *whatever* they may be. Those inoffensive brackets serve to collect the members.

There is next an anonymous but associative operation on G, so that whatever the a and whatever the b, that operation leads inexorably to some object c, one that is already within the group and so a member of the team. Symbols convey the action of the operation very elegantly: $a \circ b = c$, and if they are now a luxury, those symbols are shortly to become a necessity.

There is in addition an element e in the group serving as an identity, so that for every a in G, the identity returns every element to itself: $a \circ e = a$.

And finally for every element a in G, there is an inverse a^{-1} taking a to the identity e, so that $a \circ a^{-1} = e$.

For reasons that I cannot explain, I have always found it helpful to imagine these rather formal statements being uttered by a stout and somewhat disheveled Mexican army officer in one of those movies of the late 1940s in which the Alamo was forever about to be overrun.

Huy señor, there is thees collective. They have thees guns bah only one operation. . . .

Algebra has this effect.

Like many other highly structured objects, groups have parts, and in particular they may well have subgroups as parts, one group nested within a large group, kangarette to kangaroo. The even integers are, for example, a group in their own right; and they are as well a subgroup of the group comprising all the integers. Yet like clients in a divorce lawyer's office, not all subgroups of a given group are alike. Some are simply weird. It is the concept of a *normal* or an *invariant* subgroup that I am after, and although the discussion that follows is labeled a *definition*, this conveys an entirely misleading suggestion of something conventional. A definition in mathematics is an exercise in uncovering the essence of things, one reason that good definitions are so hard to pull off, since a definition brings the essence to light, and the light brings the definition to life.

The group is once again the very specific **G**—the integers under addition. I am now going to construct a specific subgroup **H** of **G**. *Do what it takes* is the mathematician's watchword (along with *take what is done,* of course). Admission to **H** is a matter of multiplication by seven. Thus while

$$\mathbf{G} = \{ \ldots -4, -3, -2, -1, 0, 1, 2, 3, 4 \ldots \},$$

$$\mathbf{H} = \{ \ldots -28, -21, -14, -7, 0, 7, 14, 21 \ldots \},$$

with **H** both a part of **G**, an aspect of the whole, and an object in its own right, and so a chip off the old block.

H is an ordinary subgroup in the sense that its identity is stable, and the problem before us is to identify the source of its stability. An example points the way to a definition, which in turn points the way to the idea.

Consider an integer *within* **H**, 14, say, and any integer *within* **G**, such as 17. The sum of $17 + 14 - 17$ is again 14, and 14 is the point from which we started. The concourse between the subgroup and its enveloping group leaves the *subgroup* unchanged. And this is, of course, just what is meant by a stable identity, whether in mathematics or in daily life.

At once, a definition, the last in what has been a difficult series. A subgroup **H** of a given group **G** is *normal* just in case for every h in **H**, and any a in **G**, the operation $a \circ h \circ a^{-1}$ always returns to **H**, the definition saying no more than $17 + 14 - 17 = 14$, but saying it generally, and saying it at once.

This definition is concise, and it is abstract, and for that reason strange.

But I know what you mean. By all means relax. Have a cigarette.

Mathematicians such as Ruffini, Lagrange, Legendre, Abel, Monge, and Cauchy had all been intrigued by quintic equations; they had a sense, those immensely clever men, that some simple scheme divided the equations that could be solved by the method of radicals from all the rest. Abel did in fact demonstrate that there were quintic equations that could *not* be solved by this method, but the deep connection between equations, the number five, and abstract structures in algebra eluded him. With the last hours, minutes, and seconds of his life seeping away, Galois divided his attention between equations and groups, showing ultimately that equations, like the numbers themselves, had a hidden algebraic identity, a certain structure, an inner life. They corresponded to, they embodied—*no*, they *were*—a group.

The fourth-degree equation $x^4 + px^2 + q = 0$ sings out that when something—*who knows what?*—is multiplied by itself four times, and *then* multiplied by some number p *after* being squared, and that product *then* added to its fourth power, and the result *then* added to some number q, the result is zero. The prolix prose that I have just draped over the equation's ten tight symbols serves, if nothing else, to demonstrate just why mathematicians use symbols in the first place. The letters p and q are the equation's coefficients: They stand for numbers; they are fixed for the life of the equation. They can be added to other numbers, divided, multiplied, subtracted, and their roots may be extracted; they take a good deal of rough treatment.

But whatever p and q, their identity is of no interest. It is x that the mathematician is after. Now as it happens, the equation $x^4 + px^2 + q = 0$ has four roots, or solutions, x_1, x_2, x_3, and x_4, and these roots may be expressed explicitly in terms of the equation's coefficients.

$$x_1 = \sqrt{\frac{-p + \sqrt{p^2 - 4q}}{2}}, \qquad x_2 = -\sqrt{\frac{-p + \sqrt{p^2 - 4q}}{2}},$$

$$x_3 = \sqrt{\frac{-p - \sqrt{p^2 - 4q}}{2}}, \qquad x_4 = -\sqrt{\frac{-p - \sqrt{p^2 - 4q}}{2}}.$$

Fourth-degree equations *are* solvable by the method of radicals. A formula suffices. That formula was available to the mathematicians of the eighteenth century. Each of these roots makes the equation $x^4 + px^2 + q = 0$ *true*, never a bad thing in mathematics, and once the coefficients p and q are specified, determining those four roots becomes a simple mechanical exercise.

The equation $x^4 + px^2 + q = 0$ has four roots, and thus four solutions. Four solutions, and thus four *numbers*. . . .

That Pythagorean note. Again and again and again.

Three soldiers are on parade. Dressed in red, they face the reader from somewhere east of Eden, the line of their beaver hats descending from the tallest to the shortest. Each change undertaken by these soldiers represents a permutation of their original position, one that can easily be tracked from the mud outpost walls by mathematicians keeping tabs on those bobbing beavers and assigned to the fort as imperial observers. Where originally they represented a 1, 2, 3 order of height, now, after the first command, they fall into a 132 position, Beaver 3 of old sandwiched between the bigger and lesser Beavers and falling in their shadow. Permutations are a part of a family of mental exercises that arise from the pleased recognition that things in nature are distinct and so can be put in different order.

What holds for soldiers holds, of course, for numbers, with 1, 2, and 3 admitting precisely six rearrangements or transformations:

$$e\ 123 \rightarrow 123$$

$$a\ 123 \rightarrow 132$$

$$b\ 123 \rightarrow 213$$

$$c\ 123 \rightarrow 231$$

$$p\ 123 \rightarrow 312$$

$$q\ 123 \rightarrow 321$$

The numbers in this transformation are now going to be demoted in favor of the transformations themselves, the mathematician clambering up a ladder he proposes to kick away at his earliest possible convenience. With the ladder, and those soldiers gone, the six transformations—*the transformations themselves*—reveal themselves to be a group. The language in which this idea is expressed may well be abstract, but the idea simply captures the activities taking place on the parade ground just noted. When the obviously gay but monstrously conflicted lieutenant complains petulantly about the sloppy way in which the men have lined up, he is embedding in language the reality of such things in the world as *lining ups, falling ins,* and *falling outs;* the permutations are nothing more than the abstraction behind various parade grounds.

It remains for us to complete the definitional tableau by verifying that these permutations comprise a group. There are four steps.

- The group's elements are the six transformations *e, a, b, c, p, q.*
- The group's operations comprise one transformation followed by another, as when *a* o *b* takes 123 first to 132, and then to 231.
- The group's identity is the permutation that does nothing—*e.*
- The group's inversion is the permutation that for each permutation goes back to the identity.

Three numbers have given way to six transformations, and six transformations have given way in turn to one group. What holds for three numbers holds for four numbers, of course, and for more than four numbers as well. The result in each case is called the symmetric group on *n*-letters.

Like any living object, the symmetric groups display a remarkable variety, and they often have a good deal of internal structure. In the case of the symmetric group on three letters, at least two smaller groups lie hidden in the group itself, from which they may be extracted by means of mathematical dissection. Peeling away and discarding the transformations *a*, *b*, and *c*, the mathematician is left with *e*, *p*, and *q*. This, too, is a group, as a definitional check quickly confirms. But then so is the identity element *e*. A group to the core, it satisfies the definition of a group all by itself.

What is suggestive in this sequence of groups within groups is the appearance and reappearance of *numbers*. The three numbers 1, 2, and 3 have gone over to a group of six transformations {*e*, *a*, *b*, *c*, *p*, *q*}. The group that results has just now been dissected into two subgroups, {*e*, *p*, *q*} and {*e*}, both of them normal and so stable. Let us now *count* the number of elements in each group. The group itself has six members; thereafter the subgroups have, respectively, three members and one member.

Such is their order.

Let us then *divide* the order of a given group by the order of the next largest group. The numbers that result are two and three.

Such is their index.

Order and index, the words suggesting both a librarian's command *and* an antiquarian's bookstore, serve the additional purpose of prompting that fond, familiar Pythagorean nerve eagerly to twitch, for at the end of this little exercise in ordering and indexing, *prime numbers* have popped up.

Groups that admit this sort of dissection, with normal groups nested in groups of prime index, as mathematicians say, are solvable, the terminology itself suggesting the far-reaching and dramatic coordination that Galois achieved between equations and their solutions, and groups and their subgroups.

A return to the world of equations is now obligatory. If there are four roots to the equation $x^4 + px^2 + q = 0$, these four roots may be permuted in twenty-four ways. Four distinct objects may be permuted in *n*! or $4 \times 3 \times 2 \times 1$ ways. These permutations comprise the symmetric group on four letters. It is this group that corresponds to the equation itself in the very largest sense.

Now it is obvious that just as minus five and plus five when added come to zero, so, too, $x_1 + x_2 = 0$, and $x_3 + x_4 = 0$. Although this trifle has been expressed as a fact about the roots of an equation, it is equally a fact about its coefficients p and q. To say that $x_1 + x_2 = 0$ is just to say, after all, that

$$\sqrt{\frac{-p + \sqrt{p^2 - 4q}}{2}} + \left(-\sqrt{\frac{-p + \sqrt{p^2 - 4q}}{2}}\right) = 0.$$

No very complicated calculations are necessary, although some readers—*not you, of course*—may wish to take my word for it.

But only *some* permutations of the original group of twenty-four respect this constraint. While $x_1 + x_2 = 0$, not so $x_3 + x_2$. With twenty-four soldiers lined up in a row, only some rearrangements preserve the order of their height. Others do not. It is this idea of a permutation subject to a constraint that Galois appropriated.

There are, in fact, only eight permutations of the original twenty-four meeting the constraint that $x_1 + x_2 = 0$, and $x_3 + x_4 = 0$. And they, too, form a group, a subgroup of the whole, in fact. The order of Symmetric Group and Subgroup is 24 and 8, and the index is thus 3.

Galois now repeated the procedure and by the same mental motion. A new constraint comes to govern the permutations. And with it a new subgroup appears, since only *four* of the eight permutations satisfying the first constraint satisfy the second.

The procedure is repeated until only the identity is left. The order and index of groups and their subgroups is as follows:

$$\mathbf{G} \supset \mathbf{SG} \supset \mathbf{SG} \supset \mathbf{SG} \supset \{e\}$$

G	SG	SG	SG	{e}
24	8	4	2	1
	3	2	2	2.

This is a solvable sequence of groups. Subgroups are normal and there are prime numbers where they are supposed to be. That horseshoe, by the way, signifies inclusion. Once Galois realized that, at their heart, equations constitute a group, he could do what mathematicians always wish to do, and that is to use the simplicity of one idea to cut away the complexity of another. The cutting away involves a withdrawal of attention *from* an equation, with its uninformative symbols, and its redirection *to* a group.

When all this was seen clearly—*he had no time*—the sands parted to reveal a clear dividing line between solvable and insolvable equations, the line there all along but never seen before.

Everything is still hanging in the air, it is still the evening before his death, and everything is going to proceed as it has been fated to proceed; but everything has changed as well.

We may follow Galois as the first light of dawn is breaking, and after that adieu. A series of nested subgroups is solvable if and only if it is comprised of normal subgroups whose index is prime. This is a fact about groups and their parts. The example just given of a fourth-order equation suggested to Galois the dramatic far-reaching hypothesis by which he evaded death and achieved immortality. *An equation is solvable by radicals if and only if its associated group contains a solvable series of subgroups.*

The final step in what is in fact a magnificent intellectual drama now begins. It involves the demonstration that, as suspected, nature draws a careful distinction between the numbers four and five. An equation in which an unknown is taken to the fifth power has *five* roots, and so may be represented by a symmetric group whose order or size is 5! or $5 \times 4 \times 3 \times 2 \times 1$. In this there is no mystery. As it happens, the next largest subgroup in the symmetric group has order $n!/2$. The only remaining subgroup is the identity. For the symmetric group of order 5, subgrouping leads to the series

$$\mathbf{G} \supset \mathbf{SG} \supset \{e\}$$
$$120 \quad\ 60 \quad\ 1$$
$$2 \quad\ 60,$$

with that last index number—60—standing out as defiantly nonprime.

Galois, by appealing to groups and their subgroups, had discovered why certain equations were solvable by radicals and others not. But if solvability by radicals depends on solvability by groups, solvability by groups depends in turn on the properties of certain numbers, thus once again sounding that ancient Pythagorean note that number is the measure of all things.

With the coming of dawn, Galois' life in thought was at an end. He prepared himself to die and died. His last night's work—*the testament*—he entrusted to his friend, Auguste Chevalier; the manuscript, ink-blotted and suffused with passion, still survives pressed under glass, an object of veneration among mathematicians. If the odious Pescheux d'Herbinville had second thoughts about his role in mathematics, he kept them to himself. Stéphanie Félice Poterine du Motel withdrew from history and disappeared. In 1846, the French mathematician Joseph Liouville published an edited version of Galois' last work in the *Journal de mathématiques,* homage deferred and so homage denied. It was only in 1870—almost forty years after Galois' duel!—that Camille Jourdan published a reasonably accurate and complete account of Galois' theories in his treatise on algebra, *Traité des substitutions et des équations algébriques.* Galois' work entered fully into scientific consciousness at what is virtually the beginning of the modern era, just ten years before the birth of Albert Einstein. When in 1907 Hermann Minkowski recast Einstein's theory of special relativity so that it made mathematical sense, he expressed the fusion of space and time that Einstein had foreseen in the language of groups.

The publication of Galois' ideas allowed mathematicians to see that a system of architecture lay exposed beneath the shifting surface of the numbers themselves—not only groups, but semigroups, simple groups, semisimple groups, Abelian groups, Lie groups, and beyond the groups, rings, fields, lattices, and ideals. Undergoing development at roughly the same magical moment in the development of human thought, the ideas of inorganic chemistry allowed the chemist to see a remarkable new world beneath the world in which various foul-smelling chemicals sputtered in any number of Central European test tubes and flasks. It was, that world, highly organized, with inorganic matter arranging itself into families, and exchanges between and among families regulated by very definite rules of combination and association, chemical equations balanced by an appeal to chemical structures, and chemical structures determined by a few elegant and simple principles. If inorganic chemistry revealed a stable world beneath a world, algebra did as much and it did it in precisely the same way. The standard history of nineteenth-century science places mathematical physics and mathematical analysis at the forefront; in the *counter*-history of the nineteenth century, it is inorganic chemistry and algebra that are paramount.

Following Galois, mathematicians realized that groups could be studied in themselves and for themselves. Groups have fascinating internal properties. The finite simple groups contain *no* normal subgroups; they have an especially rich character, and over the past twenty-five years, in an effort spanning a dozen countries and written up in thousands of pages, mathematicians have succeeded in classifying them completely into various families.

But like every profound mathematical idea, the concept of a group reveals something about the nature of the world that lies beyond the mathematician's symbols. In the early years of the twentieth century, Sophus Lie discovered that there are continuous as well as discrete groups, and so unearthed a stunning connection between an algebraic idea and the world in which people, planets, and protons move without interruption—*our* world.

There is thus a royal road between group theory and the most fundamental processes in nature. Some groups represent—they are reflections of—continuous rotations, things that whiz around and around smoothly. On current theories, the neutron and the proton of old—the neutron neutral in charge and the proton cheerfully positive—lose much of their identity and come to be regarded as the components of a single item, the nucleon, which, like an Indian deity, can come to earth in many shapes. When it is spinning up, a proton emerges, when spinning down, a neutron. The group named $SU(2)$ represents just what stays the same and what changes in spins of this sort. No new ideas are involved. Symmetry remains symmetry, something captured by a group.

In the early 1960s, particle physicists were confronted with a virtual zoo of new particles, unstable objects that left glowing traces of themselves in various experiments but refused to cohere into any stable pattern. A scheme of organization was needed. Murray Gell-Mann and Yuval Ne'eman both realized that SU(2) was a subgroup of a still larger group—the fabled SU(3)—and that when particles were organized by SU(3), an eightfold symmetry emerged, with families of particles neatly organized into very intuitive subgroups. When one of the physical octets specified by the group appeared to be missing a member, Gell-Mann and Ne'eman both predicted that the missing particle was there and would be found—as it was.

A great many mathematical physicists, trained originally in classical methods, were astonished by this dramatic invocation of group theory, a subject they regarded as impossibly esoteric. Mathematicians took it all in stride and wondered at the fuss, and some mathematicians even hinted broadly that had only they bothered, they would have at once seen the usefulness of SU(3). Each party proved correct in his own way. The classically trained physicists were right to be nonplussed; the mathematicians right to think they might have made the same discovery. And Gell-Mann and Ne'eman were rightest of all to regard their work as revolutionary. Curiously enough, neither the physicists nor the mathematicians found it at all surprising that a group of exotic and short-lived particles should think to organize themselves into very convenient groups.

Isolated and alone and immured in his own immature fury, it was Évariste Galois who brought this magnificent structure into being.

7

NON-EUCLIDEAN GEOMETRY

F EW MATHEMATICAL SUBJECTS SEEM quite so irresistibly lurid, the very words *non-Euclidean* suggesting an exotic universe in which embarrassing extra dimensions pop up in space and things by curving manage simultaneously to turn themselves inside out and upside down. When in 1915 Albert Einstein advanced a theory of gravity in which old-fashioned Newtonian forces vanished in favor of curved space and time, the impression was widespread that things were far weirder than anyone might have imagined. These impressions are not so much mistaken as misconceived. There is weirdness in non-Euclidean geometry, but not because of anything that geometers might say about the ordinary fond familiar world in which space is flat, angles sharp, and only curves are curved. Non-Euclidean geometry is an instrument in the enlargement of the mathematician's self-consciousness, and so comprises an episode in a long, difficult, and extended exercise in which the human mind attempts to catch sight of itself catching sight of itself, and so without end.

Mathematicians mark time by births and deaths, the greater the mathematician, the more auspicious the dates. Carl Friedrich Gauss was born in 1777 and died in 1855, his entry into and departure from the mathematical scene comprising an era all its own. Gauss is very often referred to as a prince among mathematicians, if not *the* prince, a nice turn of phrase that suggests his combination of aristocratic reticence and assured intellectual power. Respectful stories recount that when his elementary-school teacher, an odious disciplinarian as it happens, asked his charges to sum the natural numbers from one to one hundred, Gauss was able to turn down his tablet at once, the correct answer inscribed on slate, even as the dutiful donkeys in the room, chubby farm children of no intellectual distinction, scratched away industriously, adding one to two, and then to three, all the way up.

Gauss reasoned as follows: $100 + 1 = 101$. That takes care of two numbers. But $99 + 2 = 101$, as well. A pattern is at work such that when first and last numbers are added, the result is *always* 101. Now how many times must this pattern be repeated to sum all of the numbers between one and one hundred? And thereafter trailing dots will serve to mark the distinction between clever Gauss and those dutiful donkeys in the sun-filled classroom scratching away. . . .

In adolescence Gauss found himself unable to commit the full range of his thoughts to paper, so quickly and so abundantly did ideas occur to him. As a young man, he was freed from financial worry by support provided by Ferdinand, Duke of Brunswick; when the duke was mortally wounded leading Prussian troops in combat against French forces under Napoléon, Gauss was already well enough known that he was able to secure a position as a professor of astronomy at the University of Göttingen. He retained some of his inventiveness, although not all of his ardor, into his elegant old age; but there is throughout his life a sober contrast between the animal vitality of his thoughts and the careful, extremely remote, and inaccessible manner in which he presented them in public. Like a Parisian jeweler setting out the rarest of stones, he published only those of his papers that he believed had reached a state of formal perfection and lucidity.

It is a policy that I myself follow.

For more than two thousand years, Euclidean geometry had seemed to mathematicians and philosophers alike to be the very model of intellectual perfection. There it was and there it stood— the definitions, axioms, and theorems lined up in a row, the powerful, compressed, incontrovertible proofs lined up behind them; and for more than two thousand years there was, in addition, the irritating fact that whereas the first four of Euclid's postulates radi-

ated a supreme and manly self-confidence, Euclid's parallel postulate seemed somehow to make its claims diffidently, something
that every real mathematician could see and even amateurs sense.
Not *false,* surely not that, for, after all, just look: The point, the line,
the pancakelike plane all confirm the thought that through a given
point outside a given line, there *is* one and only one line parallel to
a given line. But if not false, then not *obviously* true either, or at
least not obviously true in the sense in which the other axioms of
Euclidean geometry are obviously true. If not false, and yet not obviously true, then what? It seemed possible that Euclid's parallel
postulate might be a disguised *theorem* of the system, with the full
weight of certainty displaced backward on Euclid's first four axioms. Many of the theorems in Euclid's system are *like* the parallel
postulate in the complexity of their formulation, and theorems are
in the nature of things supposed to be less evident than the axioms
from which they are derived.

Over the centuries, mathematicians thus attempted to prove
Euclid's parallel postulate, the cohort, when viewed historically,
unique in the extent to which their efforts did nothing to further
their agenda. Girolamo Saccheri, at the beginning of the eighteenth century, and Johann Lambert, some years later, both constructed ingenious arguments in which they endeavored to show
that by denying the parallel postulate they could reach a contradiction. When their work was reviewed by more patient mathematicians, it was seen that their laborious proofs in some way or
another assumed the very point at issue—the parallel postulate itself. Gauss was masterful in his discernment of self-deception
among mathematicians, noting with satisfaction in his diaries or
letters just where and just how various proofs of the parallel postulate doubled back on themselves.

By the beginning of the nineteenth century, a number of mathematicians had come to suspect that Euclid's parallel postulate
could not be demonstrated. It stood alone. Its authority was otherwise. In the *Critique of Pure Reason,* a work that Gauss read with
great diligence and no little skepticism, Immanuel Kant had ar-

gued that both space and time are pure forms of intuition, things given by the human mind and fixed by its structure. The axioms of Euclidean geometry dominate our thoughts because they cannot be dislodged from our minds. In this way, Kant lent the authority of his art to a structure about to fracture under stress. Gauss remained unpersuaded, confiding his prophetic doubts to his desk drawer, and remarking, when other mathematicians finally expressed *their* doubts about Euclidean geometry, that he had known it all along.

So far I have simply offered the outlines of a familiar mathematical fable, one suggesting nothing more than a premonitory rumble followed by its moist conceptual explosion. This is not the whole story. The discovery of non-Euclidean geometries represents a moment in *delayed* self-consciousness, with the most obvious of rumblings very often hiding the real revolutions to come.

Cackling dryly and then burying his ideas in his unpublished notes, Gauss came to suspect that Euclid's parallel postulate was *independent* of the other axioms of Euclidean geometry. The cackle is geometric, to be sure; but it is logical as well.

The first four of Euclid's axioms cover the basics; they are known as the axioms of absolute geometry:

1. To draw a straight line from any point to any point;
2. To produce a finite straight line continuously in a straight line;
3. To describe a circle with any center and distance.
4. All right angles are equal to one another.

Playfair's axiom constitutes the fifth axiom, the place where doubt arises. That axiom again:

5. Through a point outside a given line *L*, one and only one line parallel to that line may be drawn.

And its two-headed denial:

5*.   Through a point outside a given line L, there are *no* lines parallel to L

or

5**. Through a point outside a given line L, there are at least *two* lines parallel to L.

There now occurs a twitch in the great nervous system that connects all mathematicians, the living and the dead. One twitch runs from Gauss to his university friend, the mathematician Wolfgang Bolyai; earnest and plodding, Bolyai regarded the parallel postulate as an intellectual canker. "It is unbelievable," Bolyai had written, "that this stubborn darkness, this eternal eclipse, this flaw in geometry, this eternal cloud on virgin truth can be endured." Bolyai attempted one proof after the other, sending his results to Gauss by post, only to have them returned almost at once, the fateful error clearly explained.

By means of an unexplained access of intuition, Bolyai's son, Johann, gave up all efforts to prove the parallel postulate and thought instead to deny that it could be demonstrated at all. Clever boy. Placing his allegiance and then his bets on 5**, with its beckoning double star, he proposed to allow the logical consequences to flow where they might. He was enraptured. "I have created," he wrote modestly, "a new world out of nothing." His father was aghast. "You must not attempt this approach to parallels. I know this way to its very end. I have traversed this bottomless night, which extinguished all light and joy in my life. I entreat you, leave the science of parallels alone." There is, in this exchange, an appeal to a form of mathematical dread that is no longer as convincing as it might once have been.

And there at once is another twitch, this one extending some nine hundred miles to the east and touching Nicolay Ivanovich Lobachevsky, a professor of mathematics and later a successful administrator at the University of Kazan in Russia. Somewhat less melodramatic than the hysterical Bolyais, *père et fils,* Lobachevsky was a solid mathematician, and a man of Svengali-like good looks, his dark hair, straight nose, and contemptuously curved lips suggesting, of all things, a man prepared to carry on a secondary line in serial seduction in time not devoted to geometry. He took the denial of Euclid's parallel postulate in stride and without immense metaphysical anxiety.

Iss nofink.

And thereupon he got on with it, another star-crossed mathematician creating a form of non-Euclidean geometry out of something so simple as the replacement of Axiom 5 by Axiom 5**.

The result is now called hyperbolic geometry; single-starring, by way of contrast, leads to elliptic, or even to double-elliptic, geometry. In either case, the addition of Axiom 5* or 5** to Axioms 1–4 has a reverberating effect on the theorems of Euclidean geometry. Lobachevsky's reasoning in the paper that he published in *The Kazan Messenger* is simple enough to suggest his method. The initial step follows Euclid. There is a line *AB* and a point *C* lying beyond the line. And at once a radical departure from Euclid, tradition, and common sense. *All* lines through *C*, Lobachevsky affirms, may be divided into two classes. There are those that sooner or later meet *AB*, and those that do not. The lines destined never to meet *AB* are separated from the rest by a pair of boundary lines, *p* and *q*. These boundary lines, Lobachevsky affirms, are *both* parallel to *AB*. Euclid's parallel postulate has been insouciantly canceled (Figure 7.1).

Consider now, Lobachevsky argued, the angle $\pi(a)$ that is formed by the lines *Dcq* on the one side of the straight line *DC*, and by the lines *Dcp* on the other side of the same straight line. All lines forming an angle less than $\pi(a)$ will ultimately intersect the straight line *AB*; but all lines forming an angle greater than $\pi(a)$ will not.

If it happens that $\pi(a) = \pi/2$, or 90 degrees, then this version of geometry and Euclid coincide. Nothing is lost, but quite obviously nothing is gained either.

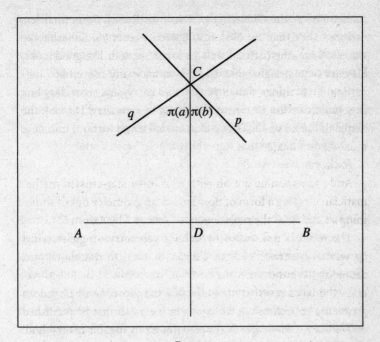

FIG. 7.1

There now follows a series comprised of *what if*'s and *what of*'s and *what iss*'s.

What if a decreases and approaches zero? In that case, π(*a*) increases and approaches π/2.

What if a increases toward infinity? In that case, π(*a*) approaches zero.

What of the sum of the interior angles of a triangle? Always less than π, but it decreases as the area of the triangle is enlarged, and approaches π as the area of the triangle shrinks.

What of the Pythagorean theorem? It is now given by the complicated formula $2(e^{c/k} + e^{-c/k}) = (e^{a/k} + e^{-a/k})(e^{b/k} + e^{-b/k})$, where *e* is the *e* of chapter 5 and *k* is a constant needed to make things come out right.

What iss circumference of circle? Iss $\pi k(e^{r/k} + e^{-r/k})$.

The conceptual adjustments to Euclidean geometry that

Lobachevsky brought about did not, of course, tell mathematicians whether they were merited and, if merited, needed. Lobachevsky and Bolyai had replaced Playfair's axiom with its denial; but they had not demonstrated that the replacement made any sense.

And they had not demonstrated that the replacement had made any sense in what we nonmathematicians consider the most fundamental sense of all. They had provided no pictures.

Did you imagine that something fancier was at issue?

Well, you were wrong.

Never mind for the moment whether Axiom 5** is *true*. The question is whether it is *consistent* with the other axioms of Euclidean geometry. And this is a logical and not a mathematical question. Now the propositions that *all whales are large* and that *all whales are mammals* are consistent in the obvious sense that no one is apt to go wrong in asserting them jointly. Not so the propositions that *all whales are mammals* and that *all whales are fish*. Since no fish are mammals, these propositions lead at once to a contradiction. In the case of whales, consistency can be easily established by a quick look at a large whale. A similar principle holds in the case of non-Euclidean geometries—a quick look suffices. To establish the consistency of non-Euclidean geometry, it is enough to provide an abstract picture, and so an imagined universe, one stripped of the sensuous essentials of the real universe and exhibiting only those features necessary to give sense to the axioms. An abstract picture of this sort is known as a model.

Now no one has ever doubted that ordinary Euclidean geometry has a model. The world in which we find ourselves is an example. But Lobachevsky's axioms have models as well, their discovery constituting the shock to the nervous system that is popularly attributed to non-Euclidean geometry as a whole. In 1866, the Italian geometer Eugenio Beltrami demonstrated that hyperbolic geometries could be modeled by the surface of a pseudosphere—the *surface*, note, and *only* the surface, so that the mathematician's imagination must serve to wrap Lobachevsky's plane around Beltrami's surface (Figure 7.2).

It can be done, the wrapping, but not completely, and when done, lines parallel to a given line start to multiply in space. But not parallel lines of old and certainly not *Euclidean* parallel lines of old. On surfaces of constant negative curvature, Euclidean straight lines undergo a semantic transmogrification; no longer straight and hardly even lines, they are identified with *geodesics,* which are arcs measuring the shortest distance between two points.

Once this adjustment to Euclid's system is countenanced, Euclid's parallel postulate unobtrusively recedes.

Lines through a point parallel to a given line?

There are lots of them.

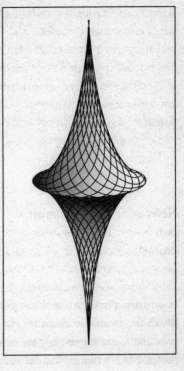

FIG. 7.2

Writing some thirty years after Lobachevsky, Henri Poincaré provided a far more intuitive model of hyperbolic geometry, one known now as the Poincaré disk. Up to a certain point, the disk is what it seems—a flat, circular, bounded, Euclidean expanse, something like a dish with no depth. Points within the disk are Euclidean points. But *lines* in the Poincaré disk consist of circular arcs intersecting the boundaries at right angles (Figure 7.3).

It is the definition of distance that changes the Poincaré disk into a model of hyperbolic geometry. Consider an arc swinging be-

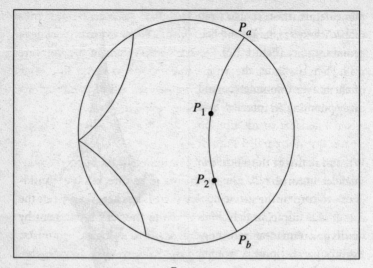

FIG. 7.3

tween two points on the circumference of the disk, *Pa* and *Pb*. Much impressed by Poincaré (instead of common sense), a snail crawling along the arc is now destined to encounter an odd phenomenon: No matter the distance covered between two points P_2 and P_1, the distance remaining between P_1 and P_a remains infinite, a mathematical illustration of the principle that the closer it seems, the farther it gets. This odd feature of space is captured by a suitably modified definition of distance, one that replaces the Euclidean definition, which simply disappears. The Poincaré distance between P_1 and P_2 with complex coordinates z and w is defined as $2 \operatorname{arctanh} \{(z - w)/(1 - wz^*)\}$, where z^* is the complex conjugate of z, and arctanh is one of those lesser trigonometric functions that are always just hanging around. This definition has the effect of forcing the boundary of the Poincaré disk forever to be an infinite distance from any point on a chord.

The Poincaré disk gains purchase on the mathematician's imagination by means of two rather considerable semantic adjustments. Lines in the Euclidean plane are reconfigured so that they become arcs; and distances in the Euclidean plane undergo a dramatic deformation so that finite Euclidean distances become infinite. With these changes in place, the first four axioms of Euclidean geometry are satisfied. But Playfair's axiom fails in the Poincaré disk. There are indefinitely many lines parallel to a given line, none of them ever intersecting the given line and all of them ingloriously pursuing an infinitely receding boundary.

What in all this of the fabled shock promised by the advent of non-Euclidean geometry? A tingle remains, to be sure, but one considerably diluted by the suspicion that there is less here than meets the eye. It is a suspicion that owes much to the very instrument by which non-Euclidean geometry has acquired its logical legitimacy. Euclid's parallel postulate fails in the Poincaré disk. Nothing beyond a model is needed to establish the point, at least to the logician's satisfaction. And yet, as a dissatisfied common sense might observe, a great deal has been accomplished simply by changing the established meaning of the Euclidean line and distance formula. Euclid's parallel lines have become swinging arcs. No wonder there are so many of them intersecting a given point. Finite distances have become infinite. No wonder that finite arcs can be called straight lines. But if the meaning of established terms can be dissolved, they can be reconstituted as well. In that case, the theorems of hyperbolic geometry become theorems of Euclidean geometry, ones that become apparent when the Euclidean line and distance formula are restored to their original state, with even Axiom 5**, turning out to be a perfectly ordinary Euclidean theorem about the property of certain arcs in the plane. If the denial of Playfair's axiom were not consistent with the absolute axioms of Euclidean geometry, the way would open to a proof of the original parallel postulate by contradiction; if any of the theorems of Euclidean geometry under their habitual interpretation were false in the Poincaré disk, then they would be false *altogether,* since disks, chords, arcs, angles, and lines are all a part of the ordinary world that reflects and so models Euclidean geometry. It is perfectly possible to scuttle backward from hyperbolic to Euclidean geometry at the least sign of trouble.

From the logical point of view, this is precisely what is wanted, establishing as it does that hyperbolic geometry is consistent *if* Euclidean geometry is consistent. But from *our* point of view, something is amiss, since the predominance of Euclidean geometry, which non-Euclidean geometry was supposed to eliminate, has simply reasserted itself like a phantom limb, one registering its presence in thought by an incessant intellectual itch and a great deal of nervous chatter.

The umbilical cord connecting non-Euclidean to Euclidean geometry was in the end severed by the work of Gauss and Bernhard Riemann. With their work, there is weirdness triumphant, a genuine upheaval in expectations and experience. If Gauss saw the way, it was Riemann who followed it. Born in 1826 in a village near Hannover, Riemann is one of the great sad figures in the history of mathematics. He was, of course, an excellent mathematician, his gifts notable from an early age; but he had in addition what can only be called a taste for the ocean floor. He got to the bottom of things. His life was a series of unrelieved misfortunes. Often indisposed, and almost always poor, he was devoted to his family, but family members died profligately and at all the wrong times; suffering from tuberculosis, Riemann died at the age of forty, his talent still in full blossom. A man of great gentleness and culture, Riemann occupies a position in mathematics similar to the one that Schubert occupies in music. They were both supremely gifted, and they were both horribly unlucky.

Riemann seems not to have known the work of Bolyai or Lobachevsky at first hand, but he had his own model of non-Euclidean geometry in the surface of an ordinary sphere. The same semantic tricks are in force. Euclid's parallel lines become great circles on the sphere, arcs whose center is the center of the sphere itself. With this readjustment of meaning, it is plain that Euclid's parallel postulate fails because there are no lines through a given point parallel to a given line. Sooner or later, great arcs all intersect. But the sphere is yet a Euclidean object and so illustrates the same now-you-see-it and now-you-don't quality as Beltrami's pseudo-sphere. Acquiring its non-Euclidean credentials by means of semantic adjustments, it surrenders those credentials the moment those adjustments are suspended (Figure 7.4).

It is for this reason that both Gauss and Riemann recognized

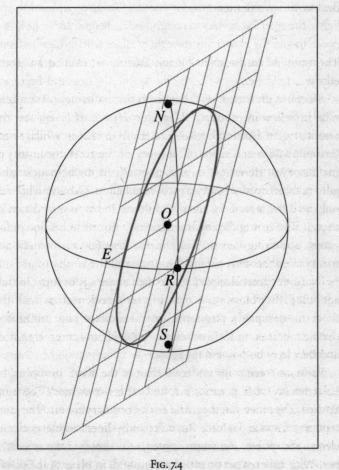

FIG. 7.4

from the first the need to create an *intrinsic* geometrical system, one that makes no assumptions about the larger world enveloping any particular non-Euclidean model. Such a system would be perfectly adapted to the snail last seen plodding along a straight line in the Poincaré disk. There is the snail, there are his local affairs, and there is what can be built from *that* and nothing else. By defining the surface of a sphere as a space in its own right, Gauss had already taken steps to accommodate that snail. With Gauss in charge, the snail loses all sense of a third spatial dimension. No up, for one thing, and no down, for another. Trudging upward by *our* lights, the snail has no way to accommodate height, and if he is disposed to give the matter any thought at all, he will register his fruitless ascent in terms of an inexplicable force resisting his every effort.

Riemann presented his thoughts to the mathematical community in a lecture entitled "Uber die Hypothesen, welche der Geometrie zu Grunde liegen—On the Hypotheses Which Form the Foundation of Geometry." It is one of the great documents in the history of mathematics, and especially happy because it contains many deep ideas and very few calculations. Gauss had himself suggested the topic for Riemann's lecture, and he was in the audience as Riemann spoke; he was afterward sincere in his praise, the young, stuttering Riemann one of the very few mathematicians who elicited the stern old man's enthusiasm.

Like every great achievement in mathematics, Riemann's lecture addressed the future while carrying on a conversation with the dead, most notably Descartes. Now Descartes had introduced mathematicians to the coordinate method, assigning a pair of numbers to every point in the plane. The coordination he achieved depends on some prior understanding of the plane, if only as the place where those points are found. This dependency Riemann severed. The plane vanishes. And so does every other intuitive conception of space. Instead, there are only the coordinates themselves, and these, Riemann argued, are simply numbers in a particular order, what he called a manifold. In place of the plane, there are sets of ordered pairs $\langle x_1, x_2 \rangle$; in place of three-dimensional space, sets of triplets $\langle x_1, x_2, x_3 \rangle$, and so to n-dimensional spaces, which are characterized by sets of n-tuples $\langle x_1, x_2, \cdots, x_n \rangle$. Numbers and only numbers, that old Pythagorean note. This is very liberating insofar as it empties the very notion of a higher-dimensional space of its mysteries. A five-dimensional space is not a strange deformation of ordinary space, one that only mathematicians can see, but a place where numbers are collected in ordered sets. When string theorists talk of the eleven dimensions required by their latest theory, they are *not* encouraging one another to search for eight otherwise familiar spatial dimensions that have somehow become lost. They are saying only that for their purposes, eleven numbers are needed to specify points. Where they are is no one's business.

With the intuitive concept of space demoted in favor of his manifolds, Riemann committed himself to the equally daring thesis that the analysis of those manifolds must be local, proceeding entirely from the properties of a manifold at a point. This is not at all the Euclidean point of view, where the Euclidean plane is given from the first. On Riemann's scheme, *nothing* is there from the first, and if nothing is there from the first, there is nothing there. The familiar image of the Euclidean mathematician observing Poincaré's snail from the outside—that also disappears. The snail is on his own.

Riemann's determination to make his analysis local made it at once possible to import the calculus into his various constructions. The derivative of a real-valued function is, after all, the supreme expression in mathematics of analysis conducted at a point—*that very one.* Riemann thus began his analysis of space by defining the distance between two points on an arbitrary manifold. There is no reason to follow Riemann all the way up to n-dimensions; two will do just fine. Descartes had defined the distance between points by means of a formula,

$$D(A, B) = \sqrt{(x_2 - x_1)^2 + (y_2 - y_1)^2},$$

one based on the Pythagorean theorem and in use since time immemorial. Riemann built his definition of distance on what are now called quadratic forms. Two points, and so two numbers, are given from the start. They are u_1 and u_2. The infinitesimal distance between those points is denoted by ds, and expressed by the formula

$$ds^2 = g_{12} \, du_1 du_2$$

The expression g_{12} on the right side of the equation is known as a tensor. It may be thought of as a general prescription, one that, like a high-school guidance counselor, indicates what kind of relationships these points may enjoy and under what circumstances. There is in this equation an astonishing amount of mathematical information, so much so that, having first announced it, Riemann provided work beyond measure for generations of mathematics. But the equation also has a limpid vernacular meaning, one made easily accessible. It specifies the distance between points, and in that sense it is simply a more general version of the Pythagorean distance formula. A geometry in which distance is defined by means of the square root of a quadratic form is known as a Riemannian geometry.

The familiar physical space in which we conduct our affairs and live our lives must now be allowed to disappear, replaced through Riemann's art by a far more abstract and inaccessible object. The connection between the ordinary concepts of geometry and concepts defined on a manifold is attenuated, but it is not severed. Thus Riemann was able to define the concept of a curve in a space of n-dimensions by means of a set of n-functions. In a two-dimensional space, two functions are required, $u_1(t)$ and $u_2(t)$, each function pegged to the parameter t, which specifies the points on the curve, point by point. With the definition of a curve in hand, Riemann was then able to define the shortest curve between two points—its geodesic; and then the angle θ between curves, angles, geodesics, and curves all arising within the manifold itself.

This part of Riemann's analysis is local, and so faithful to his original idea; but Riemann was also able to define the global properties of his manifolds, and in particular their overall curvature. The definition stretches the very margins of the imagination. The overall curvature of a sphere, say, its everlasting roundness, is not easily seen unless one simultaneously sees the enveloping space in which that sphere *is* round. Riemann provided an assessment of global curvature that makes no assumptions about the space beyond the space being studied, and so makes use of mathematical quantities entirely defined within the manifold itself. The requisite definitions Riemann took from the calculus and involve the generalization of a tangent line to a tangent plane. The sphere continues to curve as it has always curved, but it now no longer requires anything beyond itself in order to keep on curving.

The global properties of space led Riemann to understand that by adjusting various parameters in his formulas, Euclidean, elliptic, and hyperbolic spaces all arise as special cases of the more general concept of space itself, which now functions in mathematics as the remote, impalpable black background from which particular spaces emerge. Euclidean space comes about when suitable tensors are all 1 and curvature, as a result, is the same at every point. But there are radically bizarre spaces as well, places in which curvature changes at every point on the manifold. What, for that matter, is one to make of a three-dimensional spherical space, a place that is finite in extent but boundless, geodesics heading off for points unknown and then ineluctably returning to the place from which they started? All powers of visualization lapse.

With these ideas, Riemann the shy mathematician became Riemann the prophet. If there is one abstract space, and that one capable of incarnating itself in various ways, then the question of whether the space in which we live is Euclidean, elliptical, or hyperbolic, or even some unsuspected monstrosity in between, is no longer mathematical. We must seek, Riemann wrote, "the grounds of various metrical relationships outside the manifold itself, in the various binding forces which act upon it. . . ." Not yet born, it was

Albert Einstein who heard this remark in the spirit world in which he was waiting as Riemann spoke.

The debate between those who take non-Euclidean geometry in stride and those who find it the source of inexpugnable weirdness may now safely be subordinated, at least in so far as *mathematics* is concerned.

Weirdness? That remains. It remains *somewhere*. But not in space. And not in mathematics.

It is the physicists who have inherited the weird.

8

SETS

IT IS 1855 AND dried out at last, the great Gauss is about to die. And then he is gone. A long transitional era in mathematics goes with him. Cauchy, Monge, Lagrange, Legendre, Hermite, Dirichlet, Bolyai, Lobachevsky, and Bernhard Riemann have all fallen or are about to fall from the dog carts of time. Karl Weierstrass, Leopold Kronecker, Ernst Kummer, and Richard Dedekind are poised to take their place. Sober men, the German mathematicians are smoking torpedo-like Havana cigars. The era of the stolid professor has commenced, almost every important mathematician in the years between 1850 and 1900 draping his well-upholstered bottom into a university chair and from those chairs controlling access to the learned journals and the ebb and flow of graduate students and disciples. Dissertations perish in their committees. If mathematics had since the death of Euler in 1785 lost something of its untamed rhapsodic aspect, what it gained was something even more considerable—intellectual mass, soberness, organization, discipline, self-confidence. All eyes on Berlin, of course.

Let me interrupt myself to ask: Is a crack-up coming?

And to answer: Of course it is.

Georg Cantor was born in St. Petersburg in 1845 and spent the first eleven years of his life in the rich, warm, syrupy Russian milieu made accessible by his father's success as a shrewd merchant. When, in 1856, his father moved his family to Wiesbaden in Germany, Cantor acquired a second language and so a second culture, but he retained throughout his life the notably dreamy disposition of a man inhabiting a larger imaginative space than his tidy German surroundings might have suggested. Like so many other mathematicians, Cantor was enraptured as a child by mathematics, even his report cards mentioning an uncommon overall gift and a curious dexterity in trigonometry, one suggesting the reappearance of a long-hidden recessive trait. His father had thought that his son might become an engineer, a profession that like medicine occupied the middle ground between the practicalities of *Geschäft* and the rapture of mathematics. The idea filled Cantor with dismay. He won his father's permission to study mathematics, the exchange between the two men, far from being a tense domestic drama, apparently marked instead by the mutual tenderness of two romantic temperaments, both concerned to please the other.

These details might suggest a young man embarking on a modest mathematical career. Nothing of the sort. Georg Cantor initiated a great upheaval in nineteenth-century thought, carrying out one of those revolutions that like certain earthquakes survive in the form of aftershocks long after the first tremor has subsided. The effort dominated his life and it drove him to madness, so that in his last years, when not resident in various lunatic asylums, he occupied himself in proving to his satisfaction that Shakespeare had not written his own plays.

Set theory was Cantor's creation and his revolution, one that he carried out with the pained sense that he was doing a great but poorly appreciated thing. He was hardly a man suited by temperament for combat. He wished for glory, but he had no great willingness to give offense, and so he discharged the greater part of his natural aggression in resentment. His antagonist, the dapper Leopold Kronecker, regarded his work as dangerous folly and devoted himself to persecuting Cantor with an immense zestfulness. Kronecker was a professor at the University of Berlin. Elegant, able, and short, he was prepared to carry his intellectual grudges to the edge of doom. Cantor had his defenders as well as his critics, Richard Dedekind, most notably, and after Cantor had managed to alienate Dedekind, Gösta Mittag-Leffler; but despite this, he remained in a defensive crouch for most of his life, uniting in his personality the two great nineteenth-century romantic clichés of neglected genius and corrosive self-pity.

Did all of this represent anything more than a spasm of spitefulness among professors?

A great deal more, as it happened. A mathematical universe that had until then seemed perfectly adapted to a nineteenth-century drawing room was suddenly about to blow up until in the end it dwarfed even the physical universe, the one that was already oppressing astronomers with its monstrous size.

There was plenty to fight about and in a very real sense the fights have never ended, both Cantor and Kronecker thudding away at each other in that celestial ring in which combat continues long after the combatants have disappeared or turned to dust.

Set theory is unusual in that it deals with remarkably simple but apparently ineffable objects. A set is a collection, a class, an ensemble, a batch, a bunch, a lot, a troop, a tribe. To anyone incapable of grasping the concept of a set, these verbal digressions are apt to be of little help. Like points in Euclidean geometry, the sets are primitive. There are at least five kiwis among the other things in the world (the first, the second, and so on to the fifth), and directly thereafter, by means of a mental maneuver in which attention, identification, and classification are all engaged, there is that *set* of five hairy fruits. The kiwis are, of course, conceptually incidental. A set, as Cantor observed—as Cantor *insisted*—can be "*any* collection of definite, distinguishable objects of our intuition or of our intellect."

A set may contain finitely many or infinitely many members. For that matter, a set such as { } may contain no members whatsoever, its parentheses vibrating around a mathematical black hole. To the empty set is reserved the symbol ∅, the figure now in use in daily life to signify *access denied* or *don't go,* symbolic spillovers, I suppose, from its original suggestion of a canceled eye.

If sets are fundamental, so, too, membership. The number two is a member of the set of even numbers. It is at home among them. That is where it belongs—$2 \in \{2, 4, 6, \ldots\}$, the mathematician's

Greek epsilon serving to remind the reader of what he or she can in this case *see,* namely that two is right there among the even numbers. Membership is primitive. No definitions are availing. But not so inclusion, a different and less earthy concept entirely. If membership is a relationship between an individual and a set, then inclusion is a relationship between sets and still other sets, as when the set comprising the numbers two, four, six, and eight is included in the broad and noble set of even numbers. The welcome definition is obvious. $\{2, 4, 6\}$ is included in $\{2, 4, 6, 8, 10, \ldots\}$ if membership in $\{2, 4, 6\}$ guarantees membership in $\{2, 4, 6, 8, 10, \ldots\}$. As it does.

Like those wonderfully fluid galaxies that astronomers assure us meet and then merge in the night sky, sets can collide with one another, forming new sets from old. The sets $A = \{1, 3, 5, 9\}$ and $B = \{2, 4, 5, 9, 11\}$ share some members. A new set $C = \{5, 9\}$ is formed from their intersection and it is comprised entirely of elements in *both A and B.* The union of these sets is a more promiscuous affair and consists of elements in *either A or B,* all of them indiscriminately amalgamated, so that $D = \{1, 2, 3, 4, 5, 9, 11\}$. The intersection of two sets is, to continue stellar analogies, a grazing motion; their union, an undifferentiated merging.

From the first, working mathematicians turned gratefully to set theory because of its immense, its *obvious,* usefulness. It comprises a series of ideas, *and* a language, *and* a technique, its serviceability in this regard an especially ironic circumstance just because sets are not themselves obviously mathematical objects, sets serving perfectly well to collect kiwis and kangaroos, as well as numbers and points. Set theory thus seemed to nineteenth-century mathematicians to play the same role in the organization of their mental life as those stretchable, lucent plastic sacs now play in bagging fruit. The sac itself is not a fruit, but how else to collect apples, pears, and peaches without that satisfying rip, the tense tingle between thumb and forefinger, and thereafter the safely bagged pineapple swinging from the scrotum of its sac? Mathematicians who used set theory paid little attention to the fact that somewhere in Halle, a greengrocer of genius was arguing for the primacy of sacs over fruit.

In analytic geometry, points in the plane are identified with pairs of numbers. *Pairs*—meaning that the first number is first and the second, second. The underlying idea is that of things in order, but without using the very notion of things in order, the idea is by no means easy to define. Given the modest set theory already at hand, the definition emerges naturally. An ordered pair (a, b) is simply the set $\{\{a\},\{a,b\}\}$ comprising the two sets $\{a\}$ and $\{a,b\}$. This little jewel of analytical ingenuity is a way of affirming that when it comes to a and b, it is a that is first and b second rather than the reverse, but nothing in the definition appeals to the idea of order itself. Like a smile, it just appears.

Concepts that had for a very long time been half a matter of intuition and half a matter of inanition now acquire a very precise analytical voice. For more than three hundred years, mathematicians had more or less accepted the idea that a function was a rule or regularity, a mapping, a conveyance, a relationship between variables, with even simple functions such as $f(x) = x^2$ exhibiting an undignified amount of energy in taking a number—*any* number—and sending it to its square. What sending a number to another number really amounted to, mathematicians could not say, beyond from time to time suggesting that a function embodied something like a primordial act. Within set theory, functions at once lose some of their metaphysical baggage. Let us get rid of the acts and the actors. A function *is* a set of ordered pairs, so that $f(x) = x^2$ is identified with the set $\{\langle 1,1\rangle, \langle 2,4\rangle, \langle 3,9\rangle, \ldots\}$, the mathematician's dots now abbreviating those innumerable pairs of numbers in which the second is the square of the first.

And there is, finally, the way in which sets can be used to define the natural numbers, thus showing that numbers are not as primitive as Pythagoras might have thought. The definition again trades on the obvious. The number zero is identified with the empty set

∅—what else? The number one is then identified with the set that contains the empty set {∅} and so contains just one thing. The number two is identified with the set that contains just the empty set and the number one, and so on up the chain of command. In this way sets, *sets* come entirely to replace numbers. The success of this definition at once suggests the possibility that in some unspeakable way, nothing exists beyond the sets themselves.

Although Georg Cantor spent a few dutiful months at the Zurich Polytechnique, the death of his father in 1863 and the acquisition of his inheritance allowed him to move to Berlin. There he devoted himself completely to mathematics. For a time he taught high school. I have no idea why. Dressed formally in a gray cutaway and a shirt with a batwing collar, the short black tie crossed at the throat and held in place by a lustrous garnet, he must have entered the classroom as a stately professorial panda. A great crown of dark brown curls is piled on his fine, noble face, the forehead high and arched. Seated at their desks, twenty young *mädchens,* their starched shirt fronts covering their gently heaving bosoms, are twittering. This is of all things a school for girls, the daughters of Berlin businessmen, proud papas eager for their plump peach-skinned offspring to have at least some modest acquaintance with Higher Thoughts before disappearing into maternity and middle age. The young women rise in a wave and sing out, *Guten Morgen Herr Professor,* the rhythm of their respect broken by the thin piping of a few giggles.

The Panda withdraws; the young women recede.

At the University of Berlin, Cantor studied under the stern Teutonic sunlight shed by Ernst Kummer, Leopold Kronecker, and Karl Weierstrass. The great age of analysis had commenced. And for good reason. Dragging its winding-sheet from the time of the Greeks, a long series of half-hidden doubts and scruples had reappeared in various middle-European seminar and lecture rooms, and with a foul exhalation of stale denture breath began rapping a bony knuckle on all the Berlin blackboards.

From the first, Denture Breath reminded the Berlin mathematicians that the Greeks could make no sense of irrational numbers such as the square root of two. Mathematicians had since the Renaissance embraced them with the guilty sense that sooner or later they would have to explain to one another just what they were doing. Now it *was* later. The square root of two is needed, after all, to measure the diameter of a triangle whose legs are one unit in length. It thus corresponds to some distance, some tangible property of things. To say simply that there is *some* distance measured by *no* number was intolerable in itself, and intolerable physically, for those splendid electromagnetic fields that Clerk Maxwell had in 1859 introduced were continuous by definition. Containing holes where there should be numbers, no such field could be continuous, the entire fabric of physical thought, if examined too closely, likely to release a battery of nauseatingly plump moths from its redolent plush.

Sitting before the masters, Cantor absorbed not only their lectures, but their attitude—a curious combination of intellectual optimism and tense wariness. No mathematician wished to see another Bishop Berkeley arrive on the scene and with a few clever remarks demolish their pretensions to rigor. Thank God these pests were for the moment trying to make sense of Hegel's dialectic or occupying themselves with Kant's Transcendental Categories. When Richard Dedekind attempted to provide a clear, persuasive account of those damnably difficult irrational numbers, he faced an audience of mathematicians. His definition Dedekind expressed in terms of the idea of a "cut." "Every *rational* number," he observed, "effects a separation of the system into two classes . . . such that every number of the first class is less than every number of the second class." There are the numbers less than two, and the numbers greater than two, the number two cutting the number system as cleanly as a knife blade. But plainly there are cuts corresponding to no rational number whatsoever, the separation made without a knife, like a loaf of bread that divides by means of the baker's glance. There are the numbers whose square is greater than two, and those whose square is less than two. The numbers fall apart, but nothing in the numbers themselves is doing the cutting. With masterful insouciance, Dedekind brought the missing irrational number into existence by an act of will. "Wherever, then," he wrote, "we have to do with a cut produced by no rational number, we *create* a new, an *irrational* number."

There is a long pause in various seminar rooms, a cough or two, as Denture Breath shakes his cerements.

We *create* a new number?

We?

Kronecker, appearing in his usual incarnation as the Accountant of Record, if not Rectitude, thought it was all nonsense. He proposed to champion standards of mathematical probity that for the moment he alone could meet, if only because he alone wished to meet them. In an important manifesto published in 1886, he outlined his objections to various attempts to conceive [of] and establish the "irrationals" in general. The natural numbers Kronecker accepted as a God-given miracle, and he accepted them as they were given, with no embarrassing questions asked about how the *Almighty* conceived of them. A sense of skepticism that placed the natural numbers in doubt would in the end place *everything* in doubt, and, like all skeptics, Kronecker was most concerned not to allow an acid of his own devising to drip on *him*. The standard of skeptical assessment that he championed was thus carefully contrived and very cagey. Mathematicians were free to do as they wished or invent what they wanted. Who was he to scruple so long as whatever they did or invented returned to the properties of the natural numbers in a *finite* series of steps. It was, of course, a standard that no one could meet. Kronecker was terrier-bright and persistent, and the fact that he had made himself independently wealthy and lived in a splendid Berlin mansion, while it excited the envy of his enemies, did not entirely endear him to his friends.

Cantor kept quiet and lay low, devoting himself to Gauss' splendid arithmetical monograph, the *Disquisitiones arithmeticae,* and answering in his Ph.D. dissertation a question that the great Gauss has set aside. It is 1867 and he is only twenty-two.

The ontological question, W. V. O. Quine once observed, is simply this: *What is there?* The safe answer is everything, but intuitions divide on cases. In a universe consisting of five kiwis and only five kiwis, is that all there is? Or is there, in addition, the set of five kiwis, or even the Form of the Kiwi, bringing the ontological total in either case to six? Platonists in philosophy have always been eager to take that extra step beyond the kiwis; nominalists have always demurred, perhaps from a sense that in matters of being as in matters of sin, nothing exceeds like excess.

Georg Cantor was a mathematical Platonist, and more, a mathematical Plotinist, his unacknowledged master the Greek philosopher Plotinus, and the universe that he constructed from the philosophical axiom that sets, no matter their size, are as real as their members, resembles closely the overwhelming universe that in his dreams Plotinus saw.

Despite notable differences between sets containing five kiwis, five fingers, five toes, and five coughs in the humid night, all such pentavalent sets are overwhelmingly alike in one respect: They are the same *size.* This observation seems simultaneously to radiate and absorb light. That these sets contain five members means they are the same size. The light goes on. That these sets are the same size means they have five members. The light goes out. This net reduction to darkness prompted Cantor to an inspired definition. Sets are similar in size, he argued, if their members can be put into one-to-one correspondence with one another. The concept of a correspondence is one of the few mathematical ideas as comprehensible as a pane of glass. To see it is to see through it. Two sets may be placed into one-to-one correspondence when their members line up, one to one, as in those painful elementary-school dances, held in the school gymnasium, in which each girl gets one of the boys, each boy gets one of the girls, and with the engagement concluded, the pairs now plodding dutifully around the gymnasium floor, everyone has been paired off and no one left out.

It is the same thing in mathematics. It is *exactly* the same thing.

There remains a final step. To the set of sets similar in size, Cantor assigned a new number—its cardinal number. The set of all sets similar in size to five-membered sets has the cardinal number five.

In all this, the impression is strong that although words have been spun, no work has been done. This is a mistake. The most surprising surprises accumulate when infinite sets are considered, and they are up for consideration just because their claim to reality is every bit as sturdy and deserving as the claims advanced on behalf of the finite sets. If cardinal numbers could be assigned to similar finite sets, Cantor asked, could they as well be applied to similar *infinite* sets? Why not? It is thus that the first infinite cardinal number makes an appearance. It is the cardinal number of the set of natural numbers itself, and to symbolize this number Cantor chose the Hebrew alef, $\aleph_0$. It is $\aleph_0$ that designates the set of all sets that can be put into one-to-one correspondence with the natural numbers.

And thereafter to the surprises, the first a puzzle noted by Galileo. The natural numbers $1, 2, 3, \ldots$ seem intuitively more numerous than the even numbers $2, 4, 6, 8 \ldots$. An indifferent common sense might even argue that there were twice as many natural numbers as even numbers. And yet, when counted by means of Cantor's cardinal numbers, the natural numbers and the even numbers turn out to be similar. They can, after all, be put into one-to-one correspondence:

$$1, 2, 3, \cdots, n$$

$$\downarrow \ \downarrow \ \downarrow \qquad \downarrow$$

$$2, 4, 6, \cdots, 2n$$

This would seem to suggest that in respect to infinite sets, the whole is not necessarily greater than its parts. Galileo found this conclusion repugnant. It is not the stuff of common stuff. It does suggest a clash of intuitions. Cantor took it all in stride. The natural numbers and the even numbers *are* similar; they *can* be put into one-to-one correspondence; and they thus *do* share the same cardinal number. It follows that they are the same size.

So much for common sense. Whereupon common sense does what common sense always does, and retires flustered from the controversy.

When in 1869 Cantor found employment at the University of Halle, a provincial Saxonian town some fifty miles or so from Leipzig, he seemed prepared to embark on a career as a man of talent, but not genius, someone who could be expected to make a contribution without causing a commotion. Between 1874 and 1884, Cantor published an extraordinary series of papers, outlining his theory of sets and from no more than the very modest, if somewhat abstract, ideas I have already outlined, drawing the most dramatic conclusions about mathematics and the world beyond.

The appearance of Cantor's first paper on set theory coincided roughly with his marriage and the acquisition of a substantial house in Halle. In his outward life—the one that could be seen from a distance—he offered an inspired impersonation of a middle-class German academician. He taught classes, attended lectures and conferences, and participated in departmental hiring decisions, each time losing to an implacable and curiously successful Other Side. He dabbled in art with some considerable skill, a pencil sketch of his depicting a mangy dog investigating attentively its crotch, a minor masterpiece. He listened to music during chamber music evenings and played the violin, a number of plangent melodies suggesting by the cliché of their Slavic soulfulness the contrast between Cantor's life and his aspirations. In his inner life, which he conveyed to his letters, papers, and journals, he raged and brooded as he clambered over an intellectual landscape that included philosophy, mathematics, and theology, dissatisfied that where he had expected to hear praise booming up from the valley floor, an enthusiastic yodel ricocheting from one mathematician to another, there was Kronecker forever saying *no* in Berlin, with even sympathetic mathematicians unwilling to give him their whole-hearted assent, and throughout it all, the ascent marked by those missing yodels, he never for a moment lost confidence in the essential glory of his vision.

For a very long time, mathematicians had followed Aristotle in fashioning an uneasy compromise with the concept of the infinite. Mathematics is, of course, up to its nose in things that, like the natural numbers, go on forever. Limits require convergence through infinitely many steps. Fractions descend without end and points appear between points on the line or in space. These various processes, and things, and the drama of continuation that they embody, mathematicians argued to one another, were an example merely of the *potentially* infinite. The natural numbers do go on forever. This is quite true. But at any particular point, there are only finitely many numbers; and to say that the rest of them are infinite is only to say that the chain of numbers, although forever finite, gets larger and longer. What seemed patently unacceptable was the idea of a completed infinity, something infinitely large in virtue of its nature. As he so often did, Gauss offered an oracular pronouncement, one expressing his "horror of the actual infinite." "I protest," he protested, "against the use of infinite magnitude as something completed, which is never permissible in mathematics. Infinity is merely a way of speaking, the true meaning being a limit which certain ratios approach indefinitely close, while others are permitted to increase without restriction." This point of view, which is attributed often to Aristotle, is one in which purely verbal constructions—"approach indefinitely close" and "increases without restriction"—presuppose precisely the forbidden concept of infinity that they are meant to evade. The sequence of ratios $1/n$, after all, "approach[es] indefinitely close" to zero as n "increases without restriction," but for any given value of n, there are still *infinitely* many fractions between $1/n$ and zero and *infinitely* many numbers beyond n. Verbal shuffles do what verbal shuffles always do, but in this case, it is hard to see that they do anything at all beyond shuffling.

It is the forbidden concept of infinity that Cantor endeavored to rehabilitate, an effort that was to prove almost too successful as, with a few theorems and their proofs, Cantor managed to bring into existence a world of infinities beyond infinities, things suddenly multiplying without cease and expanding without limit.

Set theoretical parturition is a remarkably simple matter. Consider a simple set P consisting of the numbers one, two, and three. How many subsets does P contain? The question is no more devious in mathematics than it would be in political science were a political scientist to wonder how many committees can be formed from members of his remarkably small three-membered department. Here is an answer by enumeration: $\emptyset$, $\{1\}$, $\{2\}$, $\{3\}$, $\{1, 2\}$, $\{1, 3\}$, $\{2, 3\}$, $\{1, 2, 3\}$, the null set, I imagine, corresponding to the departmental chair, and the rest making up various combinations of professors.

And here is the answer by means of a formula: 2^N, where N is the cardinality or size of the original set.

And here is the point: 2^N is greater than N—*always*. What Cantor called the power set of a set is larger than the set itself.

Access to the infinite beyond the infinite has now been acquired, for by means of reasoning already established—the reader snuffling mildly two paragraphs before, missing the careful clue—$\aleph_0$ is less—it must be less—than the set of *its* subsets, which has the cardinality $2^{\aleph_0}$. There is thus a cardinal number lying beyond $\aleph_0$, and, of course, by means of the same reasoning, if there is one, there is another. An access to the transcendental hierarchy is now open, the cardinal numbers now lined up somewhere in space beyond the natural numbers like an endless series of freight cars:

$$\aleph_0, 2^{\aleph_0}, 2^{2^{\aleph_0}}, \ldots,$$

these admittedly strange numbers all denoting infinite magnitudes, the progression easily matched by a corresponding series of English adjectives: *big, bigger, bigger still, still bigger, bigger than bigger, humongous, and beyond. . . .*

Completed infinities? But *of course* they are completed. Their names are right there on the printed page and their existence has come about by means of the single assumption that sets are as real as the members they contain.

The hope that in potential infinities a way would be found both to have access to the infinite and to retain the comfort of purely finite magnitudes may now be seen for what it all along was, that is, an illusion. Either there are infinite magnitudes or not.

As far as mathematics goes, there had better be.

The cardinal numbers capture one property of the numbers themselves, and that is their size. But the numbers not only measure size but indicate order, so that the number ten, which the Pythagoreans worshiped, has a certain internal structure. It comes after nine, which in turn comes after eight, and so on down. In order to capture this property of the numbers, Cantor suggested that the number ten is comprised of its immediate predecessor, *and* all the numbers before it, so that $10 = 9 \cup \{0,1,2,3,4,5,6,7,8,9\}$.

Or more generally, $x^+ = x \cup \{x\}$, where x^+ is the successor to the set x. The natural numbers as a whole form a set, one completed, Cantor believed, before the mind's eye. It is the smallest set that contains zero and that contains x^+ whenever it contains x. It is designated by the symbol ω. And therein begins a drama of mathematical gestation almost biblical in its fecundity. "What happens," the mathematician Paul Halmos asked rhetorically, "if we start with ω, form its successor ω^+, then form the success of that, and proceed so on ad infinitum?" The question is sly, and it contains a double sense. What happens, one might ask, if we simply apply rules of set construction to sets already constructed? One question.

But what happens, one might also ask, as those rules of construction are endlessly applied. Another question. The first is a matter of legitimacy. Can those rules be recycled, the second a matter of ontology? Cantor believed, of course, in recycling like mad, and the very simple rule of set succession, involving nothing more than a backward look toward a set's predecessors, opens up to an enlargement of possible sets so vast as to make the natural numbers themselves seem tiny. And the natural numbers, it must not be forgotten, are *infinite*.

Thereafter Halmos is especially fine in recounting the progression of ordinal numbers in a cadence appropriate to Genesis. "After $0, 1, 2, \ldots$," he writes, "comes ω, and after $\omega, \omega + 1, \omega + 2, \ldots$ comes $\omega 2$. After $\omega 2 + 1$ (that is, the successor of $\omega 2$), comes $\omega 2 + 2$, and then $\omega 2 + 3$; next after all the terms of the sequence so begun comes $\omega 3$. Next comes $\omega 3 + 1, \omega 3 + 2, \ldots$, and after them $\omega 4$. In this way we get successively $\omega, \omega 2, \omega 3, \omega 4, \ldots$." But something follows from this sequence in the way that ω follows from $1, 2, 3, \ldots$ "That something is ω^2. After that the whole thing starts over again: $\omega^2 + 1, \omega^2 + 2, \cdots, \omega^2 + \omega, \omega^2 + \omega + 1, \omega^2 + \omega + 2, \cdots, \omega^2 + \omega 2, \omega^2 + \omega 2 + 1, \omega^2 + \omega 3, \cdots, \omega^2 + \omega 4, \cdots, \omega^2 2, \cdots, \omega^2 3, \cdots, \omega^3, \cdots, \omega^4, \cdots, \omega^\omega, \ldots$"

But the tower keeps climbing, for after the last of the omegas, there is ε_0, and then $\varepsilon_0 + 1$, and $\varepsilon_0 + 2$, and then, of course, $\varepsilon_0 + \omega$, and $\varepsilon_0 + \omega 2$, and so further to ε_0 squared, and beyond that to still another set, the tower of successor sets multiplying itself endlessly, pullulating, fecund, vast, unfathomable, what Cantor called the *unermesslichen Grösse*.

And this from no more than a handful of symbols.

Although Cantor had published a number of important papers before, he published the first of his *revolutionary* papers in 1878, and he published it in *Crelle's Journal,* over the objections of its editor, Leopold Kronecker. It was in the *Mathematische Annalen,* however, that Cantor published the six papers that most completely expressed his vision of the universe of sets. He was by then an established university professor at Halle, and so a man of some prestige, but like every mathematician who had not been called to Berlin, he found his exclusion a great affront to his happiness, and schemed incessantly to obtain a position. It was apparently Leopold Kronecker who stood in his way. Cantor fell ill, his mood darkening as the result of overwork and nervous tension; he was forced to find refuge in various spas and mental hospitals, his nurses in their starched uniforms perhaps unconsciously reminding him of those students of his who had greeted him as a young man.

Set theory is a very large tableau and like only the greatest works of art, it divides everything into before and after. Open any contemporary mathematical text at random, and the theorems, proofs, and definitions are all expressed in terms of the ideas that Georg Cantor created. The tools and the techniques of set theory have so completely been adopted by the mathematical community as to become almost identified with the tools and the techniques of mathematics itself. For more than a century, the language that Cantor invented has been the standard language of mathematics throughout the world. Set theory also permeates ordinary English as well as ordinary mathematics, the words "a very large proper subset" seeming inoffensively to designate what before 1880 would have been expressed by the words "most but not all." But to speak of the *triumph* of set theory would be to assign to its creation the attributes of a myth. Cantor's set theory is inconsistent, mathematicians understanding at once that its fecundity and its inconsistency were deeply linked. There is Russell's paradox, the best known among a collection of paradoxes and the easiest to state. If sets are subject only to some principle of free construction, then what, Russell asked, of the set of all sets that are not members of themselves? There are those various sets that are not members of themselves: the set of dogs, which is not a dog, the set of mathematicians, which is not a mathematician, or the set of blondes, which is not a blonde. Then there is the set of *all* these sets. If this set is not a member of itself, then it is, by virtue of how it is defined, and if it is, then it is not, by virtue of what it includes. This is a most troubling conclusion to have reached, the more so since it involves no fancy mathematics, no complicated definitions, nothing out of the ordinary; Russell's paradox trades simply on the very most basic ideas of a set and membership in a set. The Italian mathematician Burali-Forti had already asked pointed questions about the set of all sets of ordinal numbers, and whether it, too, was an ordinal number. Cantor had discovered paradoxes himself; but he proposed no general intellectual defense of his theory adequate to its existence. The most striking mathematical theory of the nineteenth century thus entered the twentieth century corrupted at its source.

In his later years, Cantor's mind turned toward theology and metaphysics. Catholic theologians had seen in his theory some suggestion of the forbidden doctrine that the universe might be everlasting, a position incompatible either with its creation by God or with his continuing usefulness. Cantor endeavored to show that this interpretation was mistaken. He was persuaded that mathematics was not only useful to the sciences—this no one had ever doubted—but the source of insights that even mathematical physicists could not on their own discover. This view is very common among mathematicians who have not done work outside of mathematics. "One of the most important problems of set theory," he wrote, "consists of the challenge to discover the various valences or powers of sets present in all Nature." That challenge having been met, or so Cantor was persuaded, it followed that the number of elementary particles in nature must be infinite, each corresponding to a point without extension. This aspect of Cantor's work, it must be admitted, has rarely won favor among mathematical physicists.

Or among anyone else, a form of prejudice, when the matter is soberly considered, as bitterly unfair as the initial condemnation of Cantor's purely mathematical ideas by mathematicians too timid to take large chances.

By the turn of the century, Georg Cantor had nonetheless achieved at least a part of his heart's desire: He was at last admired. No one doubted that he had changed the face of mathematics. David Hilbert, the largest mathematical personality of his time, would later refer to set theory as an intellectual *paradise,* an oddly apt choice of words suggesting both a valley of fruit trees in bloom and the prospect of an unforgiving exile.

For much of his later life, Cantor was consumed by the idea that Francis Bacon had composed Shakespeare's plays and talked endlessly about the conspiracy to obscure this fact. He again fell psychologically ill in 1916, as the fires of the First World War contracted steadily around the circumference of the German Empire, and, confined once again to a mental hospital, he was for the first time in his life forced to undergo real deprivation. He grew thin and then gaunt.

His great wish was to be given permission to return to his home on the Händelstrasse in Halle, where he had lived for so many years, but for reasons that are not clear, this wish was denied, and frail and by now worn, he died on January 16, 1918, twenty-nine years after Leopold Kronecker had died in Berlin.

9

INCOMPLETENESS

N OT A CRISIS—NO, that would be too much. More like a persistent cough, with any number of distinguished mathematicians sniffling into their handkerchiefs or hawking energetically against their raised fists. It is sometime after 1889 and before 1932, an artificial era bounded on the one hand by Kronecker's death and on the other by the publication of Kurt Gödel's monograph, "On Formally Undecidable Propositions of *Principia Mathematica* and Related Systems," in the German mathematical journal *Monatshefte für Mathematik und Physik*. Throughout years of war, revolution, and civil unrest, mathematicians in Germany, France, England, Russia, and then the Soviet Union were doing what mathematicians have always done: casting for concepts, proving theorems, streamlining proofs, and at international meetings chattering amiably in time reserved from fomenting feuds.

And yet there is that cough. Having died of a bronchial infection all his own, Kronecker kept his spirit hacking for years afterward and so became a one-man epidemic. And for good reason. A number of shocks had frazzled the collective nervous system of the mathematical community. No sooner had mathematicians grown used to hyperbolic or double-elliptic surfaces in non-Euclidean geometry than something twice as strange would pop up—a Klein bottle, say, with drooping Dali-like handles or one of Poincaré's non-Euclidean Petri dishes in which under some monstrous metric finite distances had become infinite. The great nineteenth-century analysts such as Weierstrass and Dedekind had by means of infinitely wearisome definitions clarified the foundations of the calculus, but their definitions, like light shone on cobwebs, had revealed cobwebs *behind* cobwebs, strange functions, for example, that were continuous everywhere but not differentiable at all, pathological deformities, counterintuitive counterexamples. The very well known contemporary text, *Counter-Examples in Analysis,* which is today read by every graduate student, comprises a series of misleading proofs supporting theorems that are not theorems. Much of the material is drawn from the late nineteenth century. In Italy, mathematicians were busy constructing still other horrible things, curves capable of filling the whole of space in any number of mutant dimensions. And in libraries and committee rooms, mathematicians were uncovering paradoxes in set theory itself, Russell's paradox chief among them, the others turning on various bizarre forms of self-reference in which, like a smile opening into a toothless mouth, an inescapable gap in thought revealed itself.

By the early years of the twentieth century, Kronecker's spirit could, with some justification, claim credit as a prophet as much as a scold. How that short dapper little man seems to have seen all the low secret places in the mathematician's art.

In 1900 David Hilbert delivered an address entitled "Mathematical Problems" before the International Congress of Mathematicians in Paris. Although still a young man, Hilbert addressed his audience as the mathematical king that he was. Prussian-born and German-bred, he had already made striking contributions to many branches of mathematics, and there was to his published work a kind of luminousness, so much so that it was often said that when Hilbert had finished with an area of mathematics, the subject acquired the perfection and the irrelevance of a museum exhibition. In a discipline dominated by men who were often hysterical and almost always vain, Hilbert was notable for his self-control, his measured appreciation of other mathematicians, and his generous instinct for the scientific sublime. The room, participants reported, was stuffy and the weather warm. Hilbert spoke for more than an hour, his thin tenor voice reedy. A part of his talk was devoted to the general sense that as the nineteenth century recused itself in favor of the twentieth, mathematicians faced for the first time since Euclid substantial questions about the intellectual authority of their discipline and the foundation on which it rested. The endless accumulating paradoxes were like certain forms of gossip both wickedly alluring and profoundly threatening because they represented a contracting fringe of mathematical unwholesomeness, one drawing ever closer to the center of mathematical life.

It is in this atmosphere of anxiety that Hilbert proposed the second of the problems that in 1900 he was prepared to address to the future: He asked mathematicians to provide a proof that the axioms of arithmetic were consistent. Georg Cantor had twenty years before defended consistency as the single probative standard for all of mathematics, the free creations of the human mind, like homeopathic medicine, justified because they did no harm. Hilbert now demanded that this lyrical idea itself be brought under the control of a mathematical proof. It was a proof that in 1900 Hilbert was not prepared to supply. What he did offer was a hint. Talking of the development of new ideas in mathematics, Hilbert remarked quite suddenly that "to new concepts correspond necessarily new signs." By "signs" Hilbert meant something like the *numerals* used in ordinary arithmetic, or the *pictures* used in geometry. "No mathematician," Hilbert remarked mildly, "could spare these graphic formulas."

Then facing his audience with cool assurance, Hilbert added words that seem as part of the past as strawberries served on various English lawns in the summer of 1914: "We hear before us the perpetual call. There is the problem. Seek its solution. You can find it by pure reason. . . ."

In 1910, Bertrand Russell and Alfred North Whitehead published the first volume of *Principia Mathematica*. Their ambition was to demonstrate that the principles of arithmetic could be derived from the principles of pure logic. Years later, Russell wrote in moving terms of his quite desperate desire to allow his turbulent thoughts to find repose in a mathematical structure of perfect certainty. The *Principia Mathematica* is the expression of this need, Russell quite clearly believing that if the principles of logic were not certain, then nothing could be certain at all. In that case, the human mind would be utterly adrift, a prospect that Russell regarded with loathing. The *Principia Mathematica* commanded the attention of the entire mathematical community, and if Russell and Whitehead required more than three hundred pages to demonstrate that $1 + 1 = 2$, then at least mathematicians who had read that far could say that by God one and one *were* two. Curiously enough, few among them thought to ask whether it was a point that they had ever doubted.

Hilbert was deeply impressed by the *Principia;* he had always been a mathematician with a very strong sense of the architectural. However mathematical ideas arise, he believed, it is the responsibility of the mathematical sciences "to investigate the principles underlying these ideas and so to establish them upon a simple and complete system of axioms." Russell and Whitehead had provided an axiomatic structure for arithmetic in the *Principia Mathematica.* They had carried out their work with an unheard-of degree of precision and meticulous detail. Why not say that they had done what in so many respects needed to be done in order to end the corrosive sense of intellectual insecurity that by 1910 had seeped from all the seminar rooms to become a part of the general mathematical air? For just a moment, Hilbert was persuaded. He then realized what should in any case have been obvious. Russell and Whitehead had provided a magnificent defense of a position not directly under attack. No matter their imposing and often rebarbative definitions, proofs, and theorems, the *Principia* served only to ratify what few mathematicians were prepared to doubt, and it did nothing to indemnify what many mathematicians were prepared to question—the consistency of their system as a whole. What did it profit the mathematical community to establish as a theorem that $2 + 2 = 4$ if no assurances could be given that somewhere down some awful tunnel of thought there might be another theorem, as impeccable as the first, this one demonstrating that $2 + 2 = 5$?

As Hilbert came to understand, Russell and Whitehead had not demonstrated the consistency of their system because they had not recognized the importance of the question before 1910, and they were in no position to answer it afterward.

In 1919, Hilbert turned his thoughts again to the foundations of mathematics. He was now a mathematician living in the fullness of time, world-famous, but not world-weary, and consumed completely by the desire to establish once and for all the moral grandeur of mathematics as a source of certainty. Together with his collaborator, the logician Paul Bernays, he occupied himself for the next eleven years in the elaboration of what at once was called the Hilbert program.

From the first, Hilbert's thoughts turned on a distinction that experience indicates is difficult to observe and painful to enforce. It is the distinction between signs or symbols and what they signify. The word *dog* comprises three English letters; the pooch is something else. This may well seem obvious, but when the distinction is allowed to lapse, the result is often chaos, both in philosophy and mathematics.

Within mathematics, signs are typically marks on paper; Hilbert now made such signs the subject of his concerns. The *concepts* of mathematics, he acknowledged, may well be "inadequate and uncertain," but the signs by which they are expressed belong to a realm of "extra-logical and discrete objects," and these "exist intuitively as immediate experience before all thought." In this they are like any other discrete physical object, the mind grasping the numerals as readily as it might grasp a series of sudden noises in the night, or colored marbles on a tabletop, or even, I suppose, the sharp and distinct smells of strong German cheeses. "If logical inference is to be certain," Hilbert argued, "then these objects must be capable of being completely surveyed in all their parts, and their presentation, their difference, their succession . . . must exist for us immediately, intuitively, as something which cannot be reduced to something else."

In the case of the *Principia Mathematica,* or any other axiomatic system for that matter, Hilbert was now calling for an effort at transcendence, one that severed the symbols of the system from their intended meaning. With the severance complete, there remained only the symbols—a handful of primitive shapes in the case of the *Principia.* If the primitive symbols of an axiomatic system could themselves be made the object of thought, so, too, their combination into still more complicated symbols, as when the symbols "1," "2," "+," and "=" are combined to form the formula "$1 + 1 = 2$." But then, Hilbert asked, what is a *proof* if not a sequence of such symbols, as when the symbols representing the axioms of the *Principia* lead step-by-step to symbols representing the conclusion that $1 + 1 = 2$?

The study of symbols taken as symbols comprised a new mathematical discipline, one that Hilbert called *meta*mathematics, since its subject was just the apparatus of symbols used in ordinary mathematics. Cantor had discovered in the various sets and sets of sets a new universe of mathematical objects. He had enlarged the margins of what the mathematician *noticed.* Hilbert now discovered a universe that had for thousands of years been hidden in one that was old, the symbols and symbol sequences that had always been the mathematician's crutch—their conveyance to the world of mathematical objects—now becoming mathematical objects in their own right.

Metamathematics involves a difficult mental maneuver, one akin to doublethink in Orwell's *1984.* It demands that the mathematician withdraw meaning from the symbols that he is observing while at the same time remembering what on ordinary occasions those symbols mean. During the 1920s, Hilbert, it is true, very often wrote as if mathematics were a kind of game, one played with formulas, just as chess is played with wooden soldiers; it was quite natural that mathematicians such as Hermann Weyl often took him to mean that mathematics was *only* a game. But if Hilbert on occasion wrote carelessly, he was far too great a mathematician to imagine that mathematics was nothing more than an energetic attempt to shuffle around symbols on various boards of play. The symbols did double duty. They expressed the great, the all-important truths of mathematics, and in this sense they functioned as symbols always function, going beyond themselves to touch the real world. But they also embodied a solid, discrete, and physical system, and in *this* sense, they were simply objects in play, the mathematicians' particular genius a part of the general human capacity to impress the human mind on matter.

If the symbols of a mathematical system are bound to double duty, Hilbert believed, it was in their incarnation as physical shapes that they offered the mathematician his best opportunity to bring the axiomatic system of which they were a part under firm and enduring intellectual control. Questions that had before been asked about an axiomatic system could now be asked about its formal skeleton instead, and since the skeleton consisted of physical objects existing in a world of other physical objects, they might well be the subject of attention more penetrating than the "inadequate and uncertain" concepts that they expressed.

To the metamathematician thus fell certain housekeeping chores: providing a complete inventory of the system's elementary symbols, and a precise account of the way in which they could be combined. There was next the matter of specifying the way in which one formula could be derived from another, an inventory of rules of inference that mathematicians had until then accepted as a part of the background chatter. With these rules specified, proofs would then be specified as well, since from a metamathematical point of view, a proof is nothing more than a sequence of symbols, each derived from the one before, the whole resembling a procession of elephants moving trunk to tail. In this way, Hilbert argued, "mathematics becomes an inventory of provable formulas."

Hilbert now found the means to express questions that had until then been as much a matter of inarticulate anxiety as anything else. Are the axioms of the *Principia Mathematica* consistent in the sense "that a definite number of logical steps based on them can never lead to contradictions"? A proof is required, and required on the metamathematical level. And to this question, Hilbert added another, one equally important. Are the axioms of the *Principia Mathematica complete* in the sense that for any formula that could be expressed by means of its symbols, there was a proof either of the formula or its negation? Mindful of the suspicion that such metamathematical proofs, whatever they might be, would themselves be as open to skepticism as the systems they were intended to justify, Hilbert demanded as well that they be expressed in what he called "finitary terms," the proofs themselves making use only of the combinatorial character of the symbols they were addressing. No appeals to meaning; no appeals to infinite sets; nothing but a short, direct, intuitively obvious sequence of finite steps.

For a moment Kronecker's face appears on the screen of thought, and then as rapidly vanishes.

Kurt Gödel was born in the Moravian village of Brünn in 1906, his upbringing and early education uneventful enough so that biographers have been forced to attach a certain morbid significance to a childhood episode of rheumatic fever. The younger of two brothers, he may be seen peeping from a family portrait taken in 1910, a child with puffed cheeks seated restlessly between his placid dark-eyed mama and his high-browed, mustachioed father, a man eager, judging by the distracted look in his eyes, to get rid of the photographer and return to his newspaper. That older brother is destined to become a prominent Viennese radiologist; he is off to the side of the photograph and pointing hopefully to the beautiful illustrated book on which the younger Gödel is carelessly resting his forearms. Gödel was educated in German in a central European gymnasium: foreign languages, Latin, the German classics, mathematics. His early homework papers, having been carefully preserved, betray an inevitable mistake in elementary arithmetic, a mistake common in kind to the one that Einstein was said to have made, even mathematicians taking satisfaction in observing that as children great mathematicians get things wrong. His record as a Latin student reveals a far more appropriate concordance between the child and the man. It would seem he never made a mistake.

There followed a course of study at the University of Vienna. Interested originally in theoretical physics, Gödel transferred his allegiance to mathematics after two years, and then to mathematical logic, the only subject commensurate with his obsessive need for precision. He was by all accounts intellectually graceful, generous, helpful, and clear-minded, the gathering force of his genius hidden from others by a day-to-day demeanor that friends and teachers recall as unobtrusive. He had the gift of lucidity. There was an association with a younger woman, Hao Wang remarks in his memoir, someone with "intellectual aspirations." Gödel later chose his wife from the lineup of a Viennese cabaret, Adele Porkert, the future Gnädige Frau Gödel having no intellectual aspirations whatsoever, and the marriage proving a great success.

In 1929, Gödel submitted his dissertation to his teachers, Hermann Hahn and Philip Furtwängler; it contained a proof of the completeness of elementary mathematical logic. The proof represented an achievement that had eluded a good many capable mathematicians.

Gödel then adopted a curious transatlantic way of life, lecturing during the 1930s at the new Institute for Advanced Study at Princeton, and even giving a course in elementary logic at Notre Dame, although what undergraduates prepared to admire Knute Rockne might have made of the owlish Gödel is difficult to say. There are reports that in addressing his broad-shouldered, Midwestern charges, Gödel's nose never left the blackboard. Gödel left Europe very late in the 1930s, his final decision to vacate Austria prompted, so one story goes, by alarming indications that he had been found fit for military service by the Austrian army and would soon be inducted into its ranks.

Legends began to accumulate and when, in 1940, he settled permanently in Princeton, both the confident child and the intellectually graceful, generous, helpful, and clear-minded young man submerged themselves in a personality that was reclusive, fearful, morbid, and daring.

On the seventeenth of November, 1930, the *Monatshefte für Mathematik und Physik* received the text of Gödel's monograph, "On Formally Undecidable Propositions of *Principia Mathematica* and Related Systems." The paper was published the following spring. It fell like a hammer blow on the mathematical community. Within the compass of its forty pages, it demonstrated that the Hilbert program was an impossibility. The system of the *Principia* was incomplete. From within its symbols, one could always construct a proposition such that *if* the *Principia* were consistent, no proof could be found for either the proposition or its negation. Such a proposition is undecidable. What is more, the proposition is true. Gödel's incompleteness theorem thus placed in doubt the very method of proof that had since ancient times been the mathematician's indispensable intellectual instrument, the sign of his glory. But the incompleteness theorem is itself a *theorem*, Gödel proving what could not be done, and so sustaining and sabotaging the method of proof at one and the same time.

The dismay engendered by Gödel's first theorem at once engendered a shock wave all its own in the form of his second theorem. The consistency of the *Principia*, Gödel demonstrated, is hopelessly compromised. Any proof that whole-number arithmetic is consistent requires techniques of reasoning more powerful than those found in the *Principia*. The intellectual probity of arithmetic is compromised at its source, its consistency remaining both beyond doubt and beyond proof.

It is said that when Hilbert learned of Gödel's results, he was first angry and then vexed. A great many logicians were dumbfounded and some mathematicians confused. More than seventy years later, Gödel's theorems evoke another reaction entirely, and that is exhilaration.

———

Gödel's proof is singular in that it requires virtually no background in mathematics itself. But it is nonetheless unique in the intricacy of its reasoning, and although Gödel was not himself Jewish, the only tradition in which his great paper can be placed is the ancient system of Talmudic commentary.

"The formulae of a formal system," Gödel remarks, "are . . . *looked at from the outside,* finite series of basic signs." We may as well follow Gödel in listing the basic signs, if only to gain an appreciation for the level of precision that his proof demands. There are, first of all, the logical constants: "~" (not), "$\lor$" (or), "$\forall$" (all), "0" (zero), "f" (the successor of), "(," ")" (left and right parentheses). There are next individual variables, "x," "y," "z," and the like. These range over the natural numbers (as in the formula "x is a prime number"). There are, next, variables that stand for classes or sets of numbers; and there are, finally, variables that range over sets of sets of numbers, wholesale collections. A formula in the system is some grammatical combination of its basic signs.

Although this system is sparse, it gains purchase on what it does not contain by definition and repetition. There are no names for the natural numbers in the system, and so no way to express the fact that four is not zero using the ordinary Arabic numeral "4." But any natural number is definable within the system by means of repeated succession, so that $ffff(0)$ stands in for the missing four, with the formula "$\sim(ffff(0) = 0)$" saying *in* the system what from a vantage point *beyond* the system we would indicate by writing "$4 \neq 0$."

At the heart of Gödel's paper is a connection between various symbols, considered strictly as physical shapes, and the natural numbers. The connection is known as Gödel numbering and it serves to bring to this branch of mathematics the power and flexibility of Descartes' method of analytic geometry. To every symbol, and then to every formula, and then to every series of formulas, and then to every series of such series, Gödel assigned a unique number. Elementary symbols are mapped to specific prime numbers: "0" $\leftrightarrow 1$, "f" $\leftrightarrow 3$, "~" $\leftrightarrow 5$, "$\lor$" $\leftrightarrow 7$, "$\forall$" $\leftrightarrow 9$, "(" $\leftrightarrow 11$, and

")" ↔ 13; more complicated formulas and series of formulas are then mapped to more complicated prime numbers.

The reader, promoted now to a logician and so a companion in arms, must now commence an intellectual process in which certain facts are grasped even as they are suppressed. Professionals in public relations will have no difficulties. In a series of forty-six definitions, Gödel showed how, by means of his numbering scheme, formulas *in* the formal system could be used to comment *on* themselves, acquiring meaning and then losing it all at the same time.

The forty-sixth definition is the last, and it shows how a series of physical shapes—a formula of the system—acquires, when suitably read, the power to comment on itself. The "*B*" stands for the "*B*" in the German *Beweiss,* or proof. The requisite definition is

$$Bew(x) = (Ey)yBx.$$

This series of fourteen physical shapes must now be bathed in the wash of a number of facts, which must themselves be simultaneously held in memory, like different and slightly out-of-focus stereo-optical images that in the end fuse into a coherent whole.

These shapes function in the first place as a formal definition, the symbols "*Bew(x)*" *defined in terms* of the symbols "*(Ey)yBx.*" The definition has nothing to do with *meaning*—it is a way simply of allowing one set of shapes to be replaced by another, almost as if the logician were given license physically to maneuver scenes on some mental movie. With his hair tied in a ponytail and his yellow angora sweater draped carelessly over his shoulders, the logician, promoted now to *maître en scène,* spots "*Bew(x)*" on the slowly unwinding reel of this mathematical movie, whereupon he scowls and mutters *cut,* replacing the now discarded "*Bew(x)*" with "*(Ey)yBx.*"

These activities of *cut* and *replace* go down the definitional chain from the forty-sixth definition to the very first. Thus the symbols "*(Ey)yBx*" may themselves be cut and discarded on the cutting-room floor, replaced in turn by the symbols now visible in the forty-fifth definition, whose symbols are in turn defined in terms of the forty-fourth definition, and so back to the system's original apparatus—the scenes and so the shapes that are present at the start—whereupon the movie mathematician slaps his plump thigh in satisfaction.

But these symbols—*the very ones*—function as well as a real definition of the familiar kind in which words or symbols are endowed with meaning. The same movie is spinning over the sprockets of the very same hideously expensive cutting-room contraption, but the scene revealed is different somehow, and where before there was only black, white, and gray, there are now all sorts of subtle colors, the director, Pedro or Fedro, murmuring with satisfaction as those drab symbols of his first cut come to vibrate, and just look at that remarkable fuchsia!

Thus $Bew(x)$ says—it *says!*—that a *number x* is *Bew*, where, as Pedro or Fedro now reminds us, the symbols *Bew* refer to some property of the natural numbers, explaining to his gaping assistant, a recent graduate of film studies at Bryn Mawr, that *honey*, saying that x is *Bew* is just like saying that x is *prime*, and certainly we shoot it the same way.

The final cut—the director's cut—now follows by means of the ventriloquism induced by Gödel numbering. This same formula just seen making an arithmetical statement in that subtle shade of fuchsia now acquires a palette of quite hysterical reds and sobbing violets, these serving to highlight the *metamathematical* scene presently unfolding, for while $Bew(x)$ says something about the numbers, it *also* says that

x is a provable formula,

meaning that *honey* the number x is the number associated *under the code* with a provable formula, whereupon the director, lost in admiration for his own art, can mutter only that deep down it's a movie about a movie.

This is to trace the province and the pedigree of the forty-sixth definition. There are no other symbols to define.

Numbers and code, master and mastered, movie and movie maker.

With formulas of a formal system tagged by their Gödel numbers and so endowed with a double voice, Gödel's essential argument can now be paraphrased, something that Gödel undertook in the introduction to his monograph. The paraphrase departs from the much longer proof that follows, but it does convey with all of Gödel's matchless concision the movement of his ideas. The argument requires the same multiple post-modern perspectives already evident in that cutting room.

Suppose, Gödel argued, that the formulas of the *Principia* are being examined from outside the system, almost as if they were an endless series of studio stills.

The Bryn Mawr *honey* last seen crossing her long legs is now busy arranging these formulas in a list. There is the first, the second, and so to the nth formula, which she designates as $R(n)$. She is for the moment acting as the metamathematician's amanuensis, since $R(n)$ is not a formula *in* the system, but the *name* of a formula in the system, one that is itself expressed *outside* the system.

Knowing all the greasy ropes as she does, *honey* knows as well that there is a way to get *into* the system of the *Principia* from the cutting-room floor. It involves taking $R(n)$ as a guide to some specific formula in the system, and then inserting within that formula the numeral n for the variable x wherever x appears.

The techniques embodied in this way may themselves be expressed by a form of directorial shorthand, the sort of thing that Pedro or Fedro might bark when he wants *honey* to get back into the system, where, he may often be heard muttering, she really belongs. That shorthand is expressed by the metamathematician's formula $[\alpha; n]$, a general prescription for getting into things such that *whatever* the formula α, $[\alpha; n]$ is the specific formula in the system that results when the sign or shape naming *n* replaces *x* in whatever formula it is that α happens to name.

As *honey* remarks, the formula [α; *n*] is expressed outside the system, but the formula that it designates moans and mates and resonates *within* the system.

Impressed despite myself that a graduate in visual communication should have such an uncanny feel for formal logic, I am minded to add a whisper to *honey*'s seashell ears:

—If *n* happens to be four, that numeral is *ffff*(0); if α happens to be the formula that *x* is prime, [α; *n*] denotes the formula *in* the system that says that four is prime. When the symbols "is prime" are themselves replaced by symbols *in* the system, "*ffff*(0) is prime" becomes a formula *of* the system.

With the film now flapping over its sprockets, a difficult scene is now to come, and if *honey* is struggling to pay attention, you must, you careless *Redbook* readers, struggle as well.

The metamathematical director is speaking, and he is speaking from beyond the reach of any formal system, but he is, of course, speaking about a formal system. He means to define a set *K* of numbers.

A number *n* belongs to *K* just in case [*R*(*n*); *n*] is *not* provable, and since these various directors of mine tend to lapse at the most crucial moments, I will now take over the commentary myself. *R*(*n*) is one of the formulas on the master list of formulas, an expression in the metamathematician's vernacular, and, if truth be told, an outsider like Pedro or Fedro himself, but *R*(*n*) is also the name of a formula on the inside, one that becomes a very particular formula on the inside—I am almost tempted to lapse suddenly into the vocabulary of prison movies, with the *Principia* representing the Big House—when the numeral naming the number *n* replaces the variable *x* in the *n*th formula on the master list.

That set *K*? It has been defined—*we* have defined it—in terms of the metamathematical concept of provability. But with the film rewound, provability is just what the forty-sixth definition has already defined. By the miracle of Gödel numbering, it is a concept that can be expressed entirely from within the system of the *Principia* itself, and if you are, as I always am, now inclined to feel a shiver running up and down your collective spines, this is after all the only appropriate homage anyone can pay to the grandeur of great art.

From the fact that provability is defined within the system, it follows that there is a formula S *within* the *Principia*, such that [S; n] says that n belongs to K. This is an observation from outside the system, but as Gödel observes, indeed, as he proves, there is "not the slightest difficulty in actually writing out the formula S"—writing it out within the system of the *Principia*.

But then

$$S = R(q)$$

for *some* formula $R(q)$ on the list of formulas, since the list includes *all* formulas.

"We now show," Gödel affirmed laconically, "that the proposition $[R(q), q]$ is not decidable."

The entirely extraordinary moment has now arrived, stage and set, actors and directors, cutting-room floor and the cuts themselves for an eternal moment motionless. The argument then proceeds:

—The formula $R(q)$ is a formula from the metamathematician's beyond. It has in q a specific number, one marking its place on the list.

—But the formula $[R(q), q]$ names a formula of the *Principia*, one that defines the property of being unprovable within the system.

—When the numeral for q replaces the variable x in the formula, the formula says of the number that that numeral denotes—q, in fact—that it corresponds to the number of an unprovable formula on the master list.

—That unprovable formula is $[R(q), q]$ itself, which has just
been overheard saying of itself that *it* is not provable.

And what it says must, of course, be true. If the formula designated by $[R(q), q]$ were provable, then q would belong to K. This
would mean that $[R(q), q]$ is not provable, given the definition of K.

On the other hand, the negation of $[R(q), q]$ is not provable either. For suppose that it were. Then q would not belong to K. But
in that case, there would be a proof of $[R(q), q]$ after all.

It follows that neither the formula designated by $[R(q), q]$ nor
its negation is provable.

We ourselves may allow Pedro or Fedro to suffer a cut all his
own, restoring to prominence in Kurt Gödel the twenty-three-
year-old director of record.

The unpurged images of this spectacular argument recede; so, too,
the details of Gödel's first theorem. Directly, the second theorem
appears, this one dealing directly with the issue of consistency. It is
a theorem that John von Neumann noticed after Gödel had com-
municated his first theorem to various mathematicians; but when
he wrote eagerly to Gödel to convey his discovery, he learned that
Gödel had already discovered the same thing. The import of
Gödel's second theorem can be conveyed by means of only a few
strokes. The first incompleteness theorem affirms that *if* the system
of the *Principia* is consistent, *then* there exists an undecidable
proposition, one that may be expressed from within the cage of its
symbols. Now by means of the magic of Gödel numbering, and the
ancillary miracle of doublethink, the consistency of the system
may also be expressed by a formula within the system. Without
going into details, let us suppose that that formula is named by the
letters *CON*. Gödel's first theorem may thus be expressed entirely
within the system of the *Principia* by means of the play between
two formulas: *If CON then* $[R(q), q]$.

Looking at just this line of code more than seventy years ago, von Neumann and Gödel both observed, with precisely the same sensation of wonder and dismay, that if there were a proof within the system of CON, then by the two-step of elementary logic, there would be a proof as well of $[R(q), q]$. But this is just what the first theorem affirms is impossible.

The consistency of elementary arithmetic thus lies beyond the powers of the system for which it most counts, the demolition established by Gödel's theorem now complete.

With the publication of Gödel's monograph, the Hilbert program came to an end. With it ended, as well, all efforts to establish the moral grandeur of mathematics. The era that began with the publication of Gödel's monograph continues to the present day. An unexpected tentativeness is now a part of mathematical culture, that and a curious sense of liberation. If Gödel's theorem undercut the very pretensions of the axiomatic method, it also forced the mathematical community to appreciate with unaccustomed modesty the fact that the sources of mathematical knowledge are and remain mysterious.

During the 1930s, Gödel lectured at the new Institute for Advanced Study, his lectures themselves constituting both a presentation and an explanation of his work. A small cadre of professional logicians—Alonzo Church, Stephen Kleene, Barkley Rosser, W. V. O. Quine, Alan Turing—understood at once the implications of Gödel's theorem, and they entertained the conviction, rare even among mathematicians, that in understanding Gödel's theorem they were understanding a work of great art made possible by an intellect of great genius.

For almost thirty years, Gödel's theorem retained an esoteric aspect, one that many working mathematicians found baffling. Gödel's monograph was not published in English until 1961, and even during the 1960s, when I was studying logic at Princeton— Gödel's home, after all—the great theorem could only really be learned from mimeographed notes that Alonzo Church had carefully prepared and from a very useful popular account of the theorem written by Ernest Nagel and James Newmann.

This has now changed, perhaps as the result of Douglas Hofstadter's entertaining book, *Gödel, Escher, Bach*. And yet Gödel's theorem has retained its esoteric aspect, with many mathematicians regarding it as marginal to their own working concerns.

On the other hand, philosophers as well as physicists have attempted to appropriate Gödel's theorem for their own ends. The physicist Stephen Hawking has recently declared that he for one has lost faith in the prospects of a single unified theory of everything; it has apparently been Gödel's theorem, which he has been late in appreciating, that has persuaded him that *any* such system could not be complete if it were consistent.

This is useful work, to be sure, but frustrating as well, since no application of the theorem has the force, or the clarity, or carries the conviction of the proof itself, so for every intended application, a counterapplication may be found.

That Gödel's theorem is great, no one doubts, but what it *means*, no one yet knows. This is in its own way a remarkable tribute to its power.

When in 1942 Kurt Gödel presented himself for a citizenship examination in Trenton, New Jersey, he came prepared to argue the finer points of American constitutional principles with the presiding judge. On a doggish New Jersey day, the courtroom sweltering, Gödel may be found standing amid a crowd of refugees. Acting as his sponsor and his friend, Albert Einstein is at his side. Gödel is apparently eager to engage the judge, a shrewd, well-fed character, in some endless dispute over whether the stacked clauses of the Constitution make it possible for the American democracy to be overtaken by a dictatorship. As Einstein well knew, Gödel had the capacity to follow a logical argument to the very edge of doom; on this occasion he was prepared to do so.

The judge said something banal, and so something expected.

"On the contrary," Gödel interjected, a logical chain ten miles long just forming on his lips.

Only Einstein's alarmed interference in the proceedings, his meaty hand draped over Gödel's frail shoulder, succeeded in diverting Gödel from his ambitious plan to sabotage his own application through sheer contentiousness.

The judge must have taken it all in stride. Some years before, he had sworn Einstein to citizenship. Oddballs came as no surprise.

Unlike his great friend Einstein, who had acquired with American citizenship a protective layer of naturalized podge, Gödel remained lean and austere, his years at Princeton spent in disciplined isolation. In 1939, he succeeded in demonstrating that the axiom of choice and the continuum hypothesis were consistent with the Zermelo-Fraenkel axioms of set theory, and so drew a far-reaching connection between logical questions in non-Euclidean geometry and logical questions in the foundations of mathematics. Twenty years later, the American mathematician Paul Cohen demonstrated that the negation of the axiom of choice and the continuum hypothesis was also consistent with the axioms of set theory, thus establishing the existence of absolutely undecidable propositions in the very heart of mathematics. Gödel continued to work in mathematical logic after 1940 but, by his own admission, his interests turned from mathematics to philosophy. It was as a great natural philosopher that in 1948 he addressed the general theory of relativity, a field remote from his interests or area of competence. He succeeded in discovering a new and perfectly bizarre solution to Einstein's field equations, one that made time travel possible. His solutions required the galaxies to be in perpetual rotation, a matter that Gödel took with the utmost seriousness, filling endless drawers with close observational calculations of the night sky.

Kurt Gödel died in 1978 in Princeton, New Jersey, having quite literally starved himself into what medical examiners reported was a state of inanition. He had not been physically well; he had always found his own health a subject of absorbing interest, but, like all valetudinarians, he distrusted the doctors he assiduously sought and ignored their advice. He had long suffered psychological afflictions.

For all that, Kurt Gödel brought a conclusion to his life in precisely the same way that he brought a conclusion to his work: by means that were daring, precise, original, and irrefutable.

10

THE PRESENT

AND NOW IT IS now. Like the numbers, the history of mathematics has a beginning but it has no end.

What to make of the Bourbaki? Named after a French general of widely admired stupidity, the Bourbaki was founded in the 1930s by a group of French mathematicians said to be dissatisfied by the inadequacy of textbooks such as Goursat's *Traité d'analyse mathématique*. They determined to do better. A committee was formed and pedagogical improvements discussed. This is the myth. In all of French history, no mathematician of standing has ever concerned himself with the welfare of his students. The Bourbaki was founded to amuse the members of the Bourbaki.

Over the next twenty or thirty years, a number of very talented mathematicians became involved with the group: André Weil and Henri Cartan, Jean Delsartre, Jean Dieudonné, Claude Chevalley; and to a man, *ça va sans dire,* these mathematicians believed that their first order of business was to correct and if possible eliminate the work of other mathematicians. If contemptuous of others, they were abusive toward one another. "*Tu es foutu,*" the group's secretary wrote to Henri Cartan on reading the first draft of a chapter he had written.

Members of the Bourbaki were especially keen to demonstrate group solidarity by reaching satisfying global judgments about mathematics. These very often fell into a simple pattern in which Subject X (homological algebra) was declared good but Subject Y (finite groups) declared bad. That history has nicely reversed many of these judgments adds only to their piquancy. The Bourbaki was persuaded that mathematics was a single, immense set-theoretic structure. Speaking French, that immense structure had, like the Sphinx, only one eye and, despite its size, was thought capable of flight. The Bourbaki published more than forty volumes, the last (on spectral analysis) in 1983. These books did just what the Bourbaki had always hoped they would do. They got things organized. Students found the books useless, a remarkable example of their capacity for pedagogical ingratitude.

Curators and connoisseurs can spot in each decade certain very fashionable mathematical preoccupations, the place where mathematicians who wished to be known went to be noticed. In the 1940s, homological algebra, category theory, and the Artin reciprocity conjectures. In the 1950s, Morse theory and differential topology, Hassler Whitney, René Thom, and Steven Smale fixtures in all the better lecture rooms. It was on the beaches of Rio de Janeiro, Steven Smale has frequently had occasion to remark, that he settled the infinite dimensional Poincaré conjecture. Ten years later, algebraic geometry seemed to mathematicians simply to scintillate, Alexander Grothendieck dominating the field with what René Thom once described as crushing technical superiority. Grothendieck has since given up mathematics and is said to be resident in a cave somewhere in the south of France, where he is occupied by various ecological issues. The classification of the finite simple groups, I suppose, is next, and after that the Taniyama-Shimura conjecture and the proof of Fermat's famous theorem, the work collectively of Andrew Wiles, Ken Ribet, Barry Mazur, and Gerhard Frey.

But this list, resembling as it does various trite accounts of what is in and what is out, or what is hot and what is not, could easily be rewritten in a dozen different ways, evidence that mathematics no longer has what for so long it had, and that is a stable center.

If lacking a center, the modern era in mathematics nonetheless displays certain identifiable but inconsistent tendencies, almost as if a river were suddenly to separate itself into a number of hissing streams.

Mainstream mathematics has become progressively more abstract, and the very best mainstream mathematicians have adopted an imperial air of command, giving directions over entire floodplains of mathematical research. In the 1960s, the American mathematician Robert Langlands imagined that he saw a wide-ranging unification in prospect between various branches of mathematics, a hot wiring among hot topics. He made his vision manifest in the form of a number of conjectures that he first advanced in an audacious letter to André Weil. Copies of that handwritten letter now circulate on the Internet, like photographs of the Shroud of Turin. The Taniyama-Shimura conjecture is an example: It suggests the others. Two mathematical worlds are under inspection. There is, in the first place, the modular forms. These are highly symmetrical structures found in certain hyperbolic spaces. And there is in the second place the elliptical equations. These are equations of the form $y^2 = x^3 + ax^2 + bx + c$. Symmetrical structures, hyperbolic spaces, and elliptical equations suggest all the Sons of Art: Lobachevsky, wagging his finger in the Russian fashion, the doomed Galois, Descartes, and even Pythagoras, still consumed by numbers after all these years.

At some time in the 1950s, two young Japanese mathematicians, Yutaka Taniyama and Goro Shimura, conjectured that to every modular form there corresponded an elliptical equation and vice versa. There was at the time no reason whatsoever to think the conjecture true; and apart from a handful of examples, no particular reason to think it plausible. Yet in his proof of Fermat's theorem, Andrew Wiles established a restricted case of the Taniyama-Shimura conjecture, thus uncovering a profound unity between branches of mathematics that had until then seemed only distantly related, like cousins who turn out to be twins.

In his letter to Weil, Langlands had proposed a much more far-reaching program of unification, one that would embrace algebra in the form of Galois theory, analysis via the automorphic forms, and number theory by means of the representation of certain finite groups, a unification in the end reaching the ancient metaphysical ideas of identity, structure, and number. Wiles' success in establishing the Taniyama-Shimura conjecture transformed Langlands' program from a romantic quest into a pursuit, one very similar to the search for a grand unified theory in mathematical physics. If carried to completion, the program would reveal that there is only one law in all of mathematics, and that a simple statement that things thought distinct are identical. Mathematicians assign Langlands himself some of the virtues that political scientists in retrospect assign to General George C. Marshall, observing that like Marshall, Langlands has been far-reaching, visionary, bold.

And, I should add, lucky.

If there are powerful unifying forces in modern mathematics, the reverse is true as well. The American Mathematical Society lists fifty main mathematical specialties, ranging from Algebraic Topology to Zermelo-Fraenkel set theory. These fifty specialties divide into more than three thousand subspecialties. Communication across mathematical boundaries is at times alarmingly difficult. When I was a lowly teaching assistant, I once asked the senior graduate student assigned me as a mentor how to handle a textbook problem in differentiation. I had no idea. It was all Greek to me. And to my *führer* as well.

"I don't do continuous," he said amicably.

Some mathematicians have the uneasy feeling that the ecological niche they have for so long occupied is about to be overrun, another coarser, cruder form of life already nibbling at its fringes and so encroaching on its territory. "The computer," the mathematician John Casti has remarked, "changes everything." Casti's remarks were prompted by the publication of Stephen Wolfram's *A New Kind of Science*, a work devoted to the proposition that the computer does, indeed, change everything, but curiously enough, if it is true that the computer changes everything, it is also true that it has not so far changed anything. Most mathematicians do mathematics by means of the time-tested triplet of pencil, paper, and patience. And yet, as all those dinosaurs must have dimly sensed, change is coming.

The origins of the computer, considering the issue in terms of various paternity tests, lie in thought experiments conducted in the 1930s and early 1940s by a small, isolated group of mathematical logicians: Kurt Gödel, Alonzo Church, Stephen Kleene, Barkley Rosser, Alan Turing, Emil Post. Moving from the shadows to the spotlight, and standing there in some consternation, the ancient idea of an algorithm or an effective procedure seemed suddenly in need of a precise definition. The informal idea is perfectly obvious. An algorithm is a linked series of rules, a guide, an instruction manual, an adjuration, a way of getting things done, a tool to address life's chattering chaos in symbols. The pilot's checklist (wing flaps—*check;* rudder stabilizer—*check;* breath mints—*check*) embodies an algorithm; so, too, the steps required to bring charges of sexual harassment at the University of Connecticut (find a man—*check;* charge him—*check*), and even the undertaker's Thoughtfulness List (Remembrance Pillow—*check;* Eternity Slippers—*check*). Elementary arithmetic embodies a series of elementary algorithms. It can be taught in no other way. In adding twenty-seven to thirty-five, the higher functions disengage themselves. A mechanical routine takes over. *Add seven and five by calling the sum from memory. Carry the one. Add two and three by calling the sum from memory. Don't forget to carry the one. Relax, yes, of course, have another cigarette, why not?* To logicians struggling to bring the concept of an algorithm to self-consciousness, it must have seemed as if they were for the first time recognizing some feature of life so obvious as for centuries to go unnoticed. But examples suggest only what an algorithm is, not how it may be defined.

In 1931, Kurt Gödel introduced logicians to the primitive recursive functions. These are numerical functions taking inputs to outputs and then treating outputs as inputs, as when the successor function sends a number n to its successor $S(n)$, and *then* sends $S(n)$ to *its* successor $S(S(n))$, so that the primitive recursive functions resemble a snake swallowing its own tail and improbably growing larger as a result. Nonetheless, succession suffices to define the addition of two numbers a and b. If $b = 1$, then $a + b$ equals the successor of a. And so it does. If $b \neq 1$, then $a + b$ is the successor of $a + c$, where b is the successor of c. And so again it is again.

Some years after Gödel presented his results, the American logician Alonzo Church defined what he called the lambda-computable functions. And to roughly the same point since the recursive and the lambda-computable functions, although quite different, did the same thing and carried on in the same way.

In 1936, Alan Turing published the first of his papers on computability, "On Computable Numbers with an Application to the *Entscheidungsproblem*," and so gave the idea of an algorithm a vivid and unforgettable metaphor. An effective calculation is *any* calculation that could be undertaken, Turing argued, by an exceptionally simple imaginary machine, or even a *human* computer, someone who has, like a clerk in the department of motor vehicles or a college dean, been stripped of all cognitive powers and can as a result execute only a few primitive acts.

A Turing machine consists of a tape divided into squares and a reading head. Although finite, the tape may be extended in both directions. The reading head, as its name might suggest, is designed to recognize and manipulate a finite set of symbols—0 and 1, for most purposes. Beginning at an initial square, the reading head is capable of moving to the left or to the right on the tape, one square at a time, and it is capable of writing symbols on the tape or erasing the symbols that it is scanning. At any given moment, the reading head may occupy one of a number of internal states. These, too, are finite and correspond in an unspecified way to various internal configurations of the reading head, so that what the Turing machine does is a function not only of what at any moment it sees, but what at any moment it *is*.

There is not much else. A Turing machine is a deterministic device, and like an ordinary machine, it does what it must do. Turing machines can nonetheless do rather an astonishing variety of intellectual tasks. Addition is an example, as with a few notational elaborations the logician may be observed constructing a calculating device out of thin air.

Those elaborations: The machine's single symbol is 1. Any natural number n the machine represents as a string of $n + 1$ consecutive 1's, so that 0 is 1, and 1, 11, and $n + m = (n + m) + 1$. The following ten lines of code suffice to get this machine to add any two natural numbers.

Any two (Figure 10.1).

A mathematical function is computable, logicians say, if there is a Turing machine by which it can be computed. In the late 1930s, Alonzo Church and Stephen Kleene demonstrated that the recursive functions, the lambda-computable functions, and the Turing computable functions were one and the same, and so provided a single precise explication of the informal notion of an algorithm. The definition of the algorithm, as Gödel observed, has one unexpected property: It is completely stable in the sense that it does not depend on any underlying system of formalization. No matter

STATE AND SYMBOL	READING HEAD
State 1, Symbol 1	Moves right, stays in State 1
State 1, Blank	Prints 1, goes to State 2
State 2, Symbol 1	Moves right, stays in State 1
State 2, Blank	Moves left, goes to State 3
State 3, Symbol 1	Erases 1, stays in State 3
State 3, Blank	Moves left, goes to State 4
State 4, Symbol 1	Erases 1, stays in State 4
State 4, Blank	Moves left, goes to State 5
State 5, Symbol 1	Moves left, stays in State 5
State 5, Blank	HALT

FIVE-STATE TURING MACHINE CAPABLE OF ADDING TWO NUMBERS

FIG. 10.1

where logicians begin in defining the algorithm, they always end up in the same place. Gödel regarded this circumstance as a miracle.

And so it is.

In virtue of its elegance and obviousness, the definition of an algorithm in terms of Turing machines has become the standard. The usual operations of arithmetic are all Turing-computable. So, too, most algorithms. So, too, Church argued, *any* algorithm, the concept of a Turing machine exhausting completely the idea of an effective calculation. This last claim is known as Church's thesis, and in the seventy years since it was advanced, it has passed progressively from a conjecture, to a persuasive definition, to a law of nature, to an unbudgeable fixture in contemporary thought.

The algorithm is the second of two great ideas in Western science; the first is the calculus. I have said this before, but I am so pleased with the thought that I am eager to say it again. In speaking of great ideas, I mean greatness in terms of the sense, retrospectively acquired, that without the idea the world would not at all be the world we know. The advent of the algorithm as a precisely defined logical concept may not have made the digital computer inevitable, but it did make it possible, a fascinating example in the history of thought of an idea engineering its own instantiation in matter.

Still, wherein the great changes in mathematics, or even the single change that might change everything? A few mathematicians have used the computer to prove certain conjectures. The most famous example is the four-color problem. Posed originally by Francis Guthrie in 1852, and then posed again by Arthur Caley in 1878, the four-color question asks whether four colors are sufficient to color any map so that contiguous regions all receive separate colors. The four-color answer—*yes*—was not received until 1976, when Kenneth Appel and Wolfgang Haken published their proof. Their demonstration required the verification of thousands of separate geometric cases, and these, Appel and Haken affirmed, had been successfully carried out on the computer.

Mathematicians and philosophers reacted to Appel and Haken's work with some distaste. Although an obvious accomplishment, it was hardly an exercise in intellectual elegance. Proofs involving thousands of special computer cases are ugly, no matter the cases, no matter the computer. But *relying* on the computer to the extent that they did seemed somehow akin to entrusting the delivery of something precious to a well-meaning but dull-witted messenger, someone apt to stall helplessly in traffic on noticing that one address is marked Sixth Avenue and the other, the Avenue of the Americas. If that messenger could get stuck, why not the computer? And stuck in just the same way, some trifling error causing it to loop pointlessly or skip over certain cases, or just round things off in the wrong way. Nonetheless, no mathematician has suggested that Appel and Haken's proof is less certain than it would be had all the calculations been done by hand. Those calculations represent dog work, and mathematicians are notably inferior to the computer when it comes to going to the dogs. (Many mathematicians cannot, in fact, add a simple column of figures with the accuracy expected of a German greengrocer.) An uneasy feeling nonetheless persists that the method of proof has somehow been compromised.

No one has quite said why.

Benoit Mandelbrot—a distant cousin of mine, a remote family connection—is a mathematician who has immensely enriched the ordinary happiness of mankind by showing how beautiful pictures can be made simply on the computer. His images are now everywhere and are everywhere known as Mandelbrot sets. Their construction depends on recursive iteration and a computer program that can assign colors to regions of the complex plane. The procedure is of remarkable simplicity. A particular complex number c is initially chosen, the complex function $f(z) = z^2 + c$ introduced. It is a function that instructs the mathematician, and ultimately the computer, to take any complex number z, square it, and then tack on c. If $z = 0$, $f(0) = c$—obviously. Thereafter, with z still 0, f is applied to itself, yielding $c^2 + c$; and so ultimately the sequence c, $c^2 + c$, $(c^2 + c)^2 + c$, ...

Pictures are generated from this scheme by means of a coloring algorithm, which in turn is contingent on a fact.

The fact first. For certain choices of c, the sequences that result are bounded. They never leave a certain region of the complex plane—a circle, say, whose origin is zero. If z stays fixed at zero, two choices yielding bounded sequences are $c = -1$, and $c =$ the imaginary i of chapter 7. Other choices of c make for unbounded sequences, the innocent-seeming $c = 1$ leading to a sequence that escapes any bound and takes off for parts unknown.

The coloring algorithm next. Examining various choices of c, and the sequences to which they give rise, the algorithm assigns one color to choices of c that yield bounded sequences, and another

FIG. 10.2

color to choices that do not. The result is the Mandelbrot set (Figure 10.2).

There is not one—there are thousands of such sets, and each reveals a landscape that is at once familiar, disturbing, and beautiful. With these new landscapes, there are new questions as well. What, for example, is the basis of the curious fact that these landscapes are *self-similar*? When blown up, the Mandelbrot set, instead of revealing an array of elementary entities, reproduces on a smaller scale all of the essential features of the landscape as a whole, so that in looking at a Mandelbrot set (Figure 10.3), the distinction between whole and part undergoes a radical dissolution, the whole in every part, and every part in the whole, and so downward for as far downward as the computer can go. Mandelbrot sets, it would seem, are immensely complicated.

And yet Mandelbrot sets are at the same time very simple to construct. This raises an interesting question of judgment. Just who is to be in charge of the business of assessing complexity—those algorithms or our lying eyes?

More curious is the fact that although they are algorithmically

FIG. 10.3

generated, Mandelbrot sets are not quite under algorithmic control. They are not, as logicians long suspected, recursive. A set is recursively enumerable if there is an algorithm that enumerates its members, one after the other. The set of even numbers is an example. And recursive if it is recursively enumerable and its complement is recursively enumerable. The even numbers again. Recursively enumerable again, and now recursive as well because an algorithm can *also* enumerate the odd numbers. Those numbers c in the Mandelbrot set leading to bounded sequences are recursively enumerable. Not so the numbers leading to the *un*bounded sequences. No guarantees are at hand, the computer just possibly dawdling over its deliberations and coming to no conclusion for the rest of time.

Those lovely pictures, the downward descent without end, complexity from simplicity, and sets that are not recursive—surely *this* is something new in our collective experience.

Is it not?

Elsewhere, the real life of mathematics goes on, the hard, fiercely demanding disciplines—analysis, algebraic geometry, ergodic theory, probability, the theory of finite simple groups, partial differential equations, algebraic topology, combinatorics—spreading inexorably beyond the reach of any program of unification. The cathedral of thought is now very high, but it is thoroughly incoherent, with a number of spectacular steeples poking through the clouds, mathematicians like Robert Langlands offering instructions to various architects, while in the basement, perpetually under revision or under water, a group of workmen in scruffy clothes, their tattoos purple in the murky light, is putting together new revetments or hoisting ten-pound sacks over their collective shoulders. Whether on the ramparts or in the cellar, no one is quite certain what they are doing, and some men are uncertain whether they are doing anything at all.

And yet much of modern mathematics is what mathematics has always been, and that is a form of art that although forever old is forever new. The calculus is by now more than three hundred years old, and yet mathematicians such as E. Cartan looked at it with innocent eyes and in creating the theory of exterior differential forms, saw things there that no one had seen before. New disciplines arise and parts of human life that had seemed completely beyond the mathematician's art become a part of his repertoire after all. When in 1948 Claude Shannon created the modern theory of information, he was doing what mathematicians have always done and that is noticing the mathematical bones behind the everyday face of a familiar concept. The manifold of mathematics is itself unstable. Odd little subjects that had for years been the province of obscure professors suddenly seem of startling importance. While studying algebra many years ago, I once had occasion to look at my professor's monograph on differential algebra. It was his life's work. No one had heard of it, and no one, judging from the inside cover, had ever checked it from the library before. A few years later differential algebra became the rage, and a few years after that, that rage was superseded by another rage. At other times, even very distinguished mathematicians carry out this exercise in reverse. When Samuel Eilenberg decided to commit his thoughts on automata theory to a monograph, he brought to bear the most exquisite of algebraic tools, but by the time the book appeared, it was too late, the expensive delicate ship that he had meant to board sailing on without him.

Professors trudge out on rainy mornings and meet their classes. There is the sound of chalk squeaking. Books are opened and notes taken. Problems are posed and sometimes settled. Things are sometimes clear, and sometimes confused. The life behind the life of mathematics goes on.

What we scribes and scribblers on the margin of this great art can say is only that everything old was once new, and that everything new will one day be old.

As for the tingling that everyone senses but that no one can specify, we must wait and see what happens.

Even so, even we.

INDEX

~

Page numbers in *italics* refer to figures.